Bashford Dean

Fishes, Living and Fossil

An outline of their forms and probable relationships

Bashford Dean

Fishes, Living and Fossil
An outline of their forms and probable relationships

ISBN/EAN: 9783337312312

Printed in Europe, USA, Canada, Australia, Japan

Cover: Foto ©Andreas Hilbeck / pixelio.de

More available books at **www.hansebooks.com**

FISHES, LIVING AND FOSSIL

Columbia University Biological Series.

EDITED BY

HENRY FAIRFIELD OSBORN.

1. FROM THE GREEKS TO DARWIN.
   By Henry Fairfield Osborn. Sc.D Princeton.

2. AMPHIOXUS AND THE ANCESTRY OF THE VERTEBRATES.
   By Arthur Willey, B.Sc. Lond. Univ.

3. FISHES, LIVING AND FOSSIL. An Introductory Study.
   By Bashford Dean, Ph.D. Columbia.

4. THE CELL IN DEVELOPMENT AND INHERITANCE
   By Edmund B. Wilson, Ph.D. J. H. U.

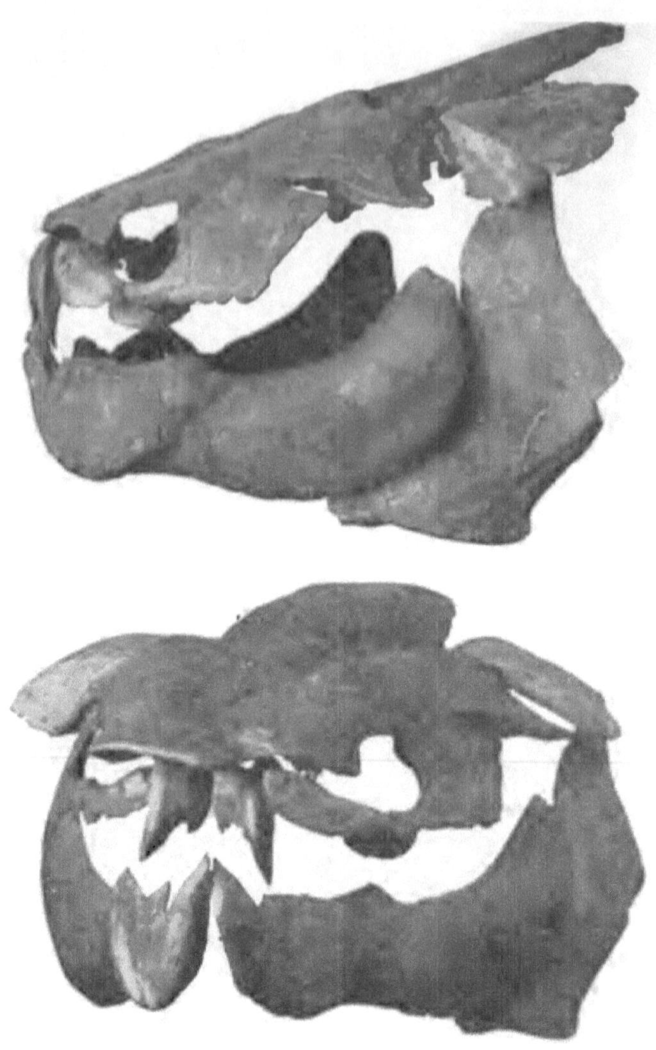

Frontispiece.—Head of DINICHTHYS INTERMEDIUS, NEWBERRY, in front and side views. × ⅒. From photograph of specimen collected by Dr. William Clark, in the Waverly (Lower Carboniferous) of Ohio, now in the collection of Columbia College, New York. (V. p. 133.)

COLUMBIA UNIVERSITY BIOLOGICAL SERIES. III.

FISHES, LIVING AND FOSSIL

AN OUTLINE OF THEIR FORMS AND PROBABLE RELATIONSHIPS

BY

BASHFORD DEAN, Ph.D.

INSTRUCTOR IN BIOLOGY, COLUMBIA COLLEGE, NEW YORK CITY

New York

MACMILLAN AND CO.

AND LONDON

1895

To

MY FRIEND AND TEACHER

JOHN STRONG NEWBERRY

LATE PROFESSOR OF GEOLOGY IN

COLUMBIA COLLEGE

Τῶν δ' ἐνύδρων ζώων τὸ τῶν ἰχθύων γένος ἓν ἀπὸ τῶν ἄλλων ἀφώρισται.

ARISTOTLE, *De Animalibus Historiae*, Lib. II., cap. xiii.

PREFACE

A KNOWLEDGE of Fishes, living and fossil, is not **to be** included readily within the limits of an introductory study. In preparing the present volume it has nevertheless been my object to enable the reader to obtain a convenient review of **the most** important forms of fishes, and of **their** structural and developmental characters. **I have also endeavoured** to keep constantly in view the problems of their **evolution.**

At the end of the book **a series of** tables affords more definite contrasts of the anatomy and embryology of the different groups of fishes.   And as an aid to further study **has** been added **a** summarized bibliography, including especially the works **of** the more recent investigators.

My sincere thanks **are** due to **my** friend and colleague, Professor Henry Fairfield Osborn, for many **suggestions** during the early preparation of the book, and for the care with which **he** has later revised **the** proof. I must also express my indebtedness to Mr. Arthur Smith Woodward of the British Museum for his personal kindnesses in aiding my studies. My thanks are also due to my father, William Dean, for the preparation of the index.

The figures, unless otherwise stated, **are from my** original pen drawings.

B. D.

BIOLOGICAL LABORATORY OF COLUMBIA COLLEGE,
May, 1895.

CONTENTS

VII

VIII

LIST OF FIGURES

I

FISHES IN GENERAL

INTRODUCTION

FISHES, defined in a popular way, are back-boned animals, gill-breathing, cold-blooded, and provided with fins. It is in their conditions of living that they have differed widely from the remaining groups of vertebrates. Aquatic life has stamped them in a common mould and has prescribed the laws which direct and limit their evolution ; it has compressed their head, trunk, and tail into a spindle-like form, it has given them an easy and rapid motion, enabling them to cleave the water like a rounded wedge. It has made their mode of movement one of undulation, causing the sides of the fish to contract rhythmically, thrusting the animal forward. A clear idea of this mode of motion is to be obtained from a series of photographs of a swimming fish (**Figs.** 1–2) taken at successive instants : thus in the case of the shark (Fig. 1) the undulation of the body may be traced from the head region backward, passing along the sides of the body, and may be seen to actually disappear at the tip of the tail. It is the pressure of the fish's body against the water enclosed in these incurved places which causes the forward movement.

The density of the living medium of fishes exerts upon them a mechanical influence ; they are, so to say, *balanced* in water, free to proceed in all planes of direction, poised

with the utmost accuracy, enabled to rise to the surface or
sink readily into deep water. A special organ, the 'air-,'
or 'swim-bladder,' has even been acquired by the majority
of living fishes, which, whatever may have been its origin
or accessory functions (v. p. 21), has certainly to an extraor-

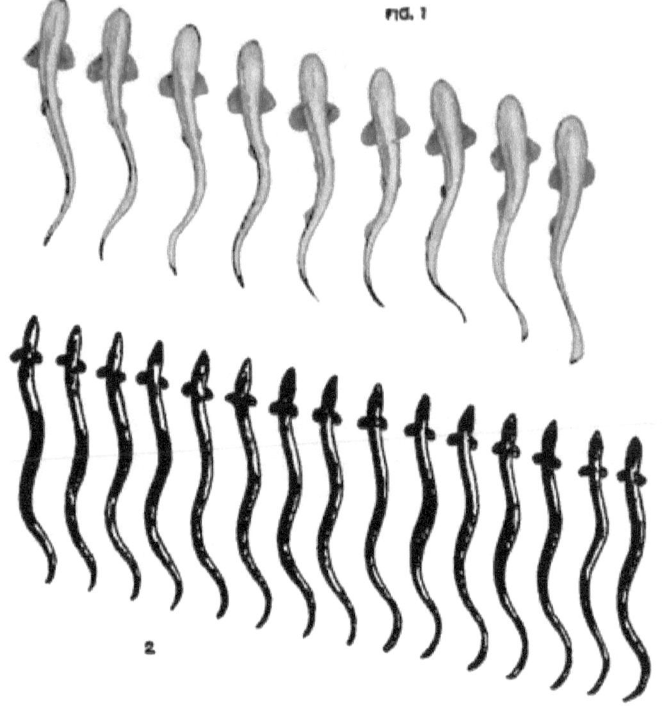

Figs. 1 and 2. — Movement of fishes, — shark and eel. (After MAREY.)

dinary degree the power of rendering the specific gravity
of the fish the same as that of the surrounding water.

In an example of a swift-swimming fish some of the
most striking peculiarities of the aquatic form may be
seen. The Spanish mackerel, *Scomberomorus* (Fig. 3),
shows admirably a stout spindle-like outline ; its entire sur-

face is accurately rounded, and there appear no irregular points which could retard the forward motion of the fish. Even in the wedge-shaped head the conical surface has been made more perfect by the tightly fitting rims of the jaws, by the smoothly closed gill shields, and by the eyes' accurate adjustment to the head's curvature. Viewed from in front (Fig. 4) the fish's outline appears as a perfect ellipse, and seems surprisingly small in size : the fins, which appear so prominent a feature in profile, can now be hardly distinguished ; above and below they form keels, sharp and thin. In side view the vertical or unpaired fins are seen surrounding the hinder region of the body : they resolve themselves into *dorsal* (*D*), *anal* (*A*), and *caudal* (*C*) elements ; the former are low and stout, elastic in their firm cutwater margin, deeply notched and inter-

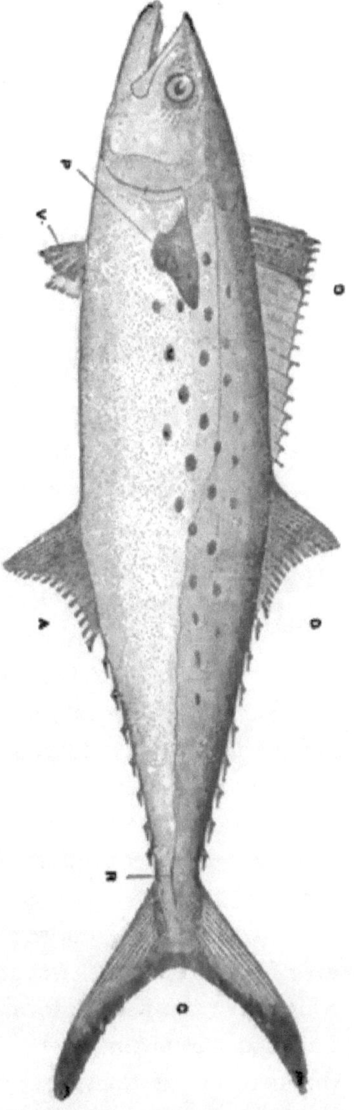

Fig. 3. — Type of swift swimming fish, Spanish mackerel, *Scomberomorus maculatus* (Mitch.), J. & G. × ¼. (After GOODE in U. S. F. C.)

rupted posteriorly, where useless elements have been discarded; the caudal is broadly forked, stout in its supporting rays, strong in power of propulsion. At its sides a remarkable ridge has been developed, functioning as a horizontal keel (*R*) and preventing the stroke of the cau-

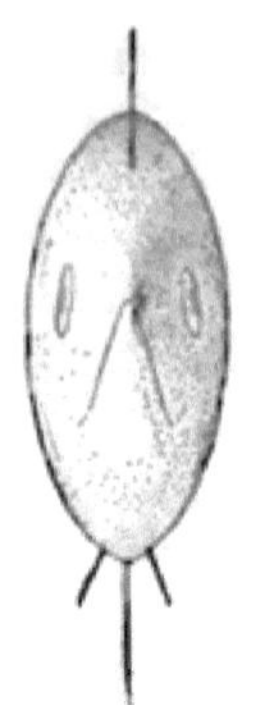

Fig. 4. — Front view of Spanish mackerel.

dal from varying from the vertical plane. The lateral, or paired fins, *pectoral* and *ventral* (*P* and *V*), may rotate outward and arrange themselves in the line of the fish's motion, so that in a somewhat horizontal plane they may, like the unpaired fins, function as keels. When thus erected, the paired fins present a firm anterior margin which serves as a cutwater. While thus somewhat similar in function to the vertical fins, the ventrals and especially the pectorals may acquire additional uses: they may serve as delicate balancers, or may aid in guiding or arresting the fish's motions.

In further conformity to aquatic needs, the entire surface of the fish is notably slime covered, and although perfectly armoured by plates and scales, yet presents no point of resistance to forward motion. An internal balance, moreover, has been effected between the supporting, visceral and muscular parts: the firm vertebral axis acquires its central position, and at its anterior end the head structures form a compact, wedge-like mass: the body muscles which give the fish its form-contour thin away on the ventral side, permitting in the region between the head and the anal fin the space occupied by the closely compacted viscera: respiratory organs occupy a restricted space on either side of the gullet; the heart and its arterial trunk are implanted closely in the throat in the median ventral line; the dorsal blood-vessel takes its position immediately below the vertebral axis, and the air-bladder in the most dorsal part of the abdominal cavity.

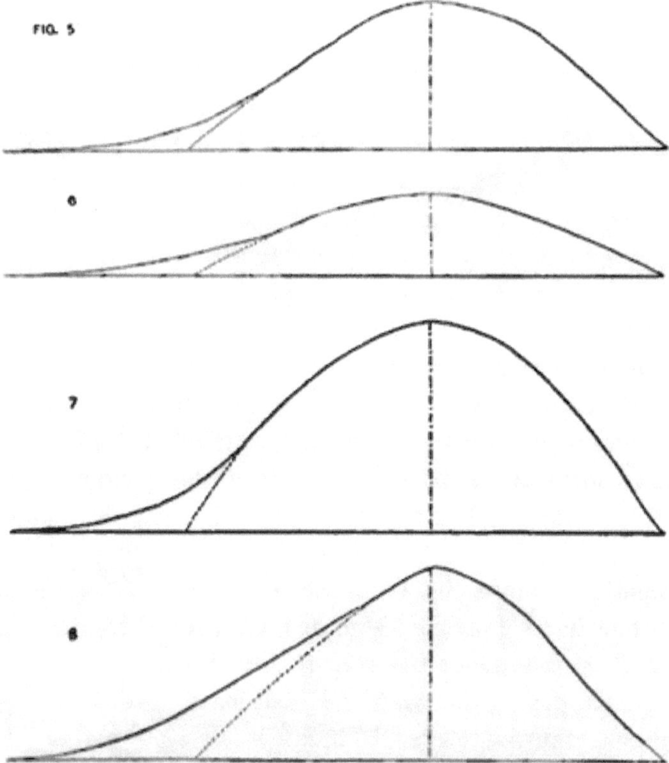

Figs. 5–8. — Numerical lines of fishes and cetaceans. The "entering angle" begins at the snout-tip at the right, and extends as far as the vertical dotted line (36 %, about, of the entire length); the "run" then begins and is continued to the body terminal. 5. Striped porpoise, *Phocaena lineata*. 6. Spanish mackerel (Cuban), *Scomberomorus cavalla*. 7. Humpback whale, *Megaptera longimana*. 8. Striped bass, *Labrax lineatus*. (All figures after PARSONS.)

In acquiring this perfect outward symmetry it is interesting to note that the forms of fishes may be said to have actually evolved the practical solution of the most theoretical problems of curves and displacement in relation to submarine motion. A study of the "lines" of typical fishes by naval engineers * has led to some most interesting results as to the uniformity of their mathematical "normals." It is found, for example, that the "entering angles" of many and very different fishes are surprisingly similar (Figs. 6 and 8) : they thus terminate regularly (at the plane of the greatest cross-section of the body) at 36 per cent of the fish's total length ; and the curves of the "run" (*i.e.* of the hinder part of the trunk, from the plane of the greatest cross-section to the body terminal), similar for all, are smooth hollow curves, which in the forward motion of the fish permit the passage of the displaced water.

It would be unreasonable to doubt that the fish form is adapted to the mechanical needs of its environment, even if there existed no further evidence than that of the metamorphoses of aquatic mammals. Many of these have shown so complete an adaptation to water-living that it is scarcely remarkable that they were early included among fishes. And it is of further interest that there exist transitional forms between the land-living mammals on the one hand and the cetaceans on the other. In the Seal it is but the initial step in the transformation that has taken place ; the head and body have become bluntly tapering, the hind legs displaced backward, the foot and hand webbed, the hair adapted to submerged locomotion. A further stage in the acquisition of the fish-like form is shown in the Dugong and Manatee. And finally in the Dolphin and Whale (Figs. 5 and 7) have been actually attained the *numerical lines* of fishes (cf. Figs. 6 and 8). In these cases, the mechanical conditions of aquatic living have produced their result only at the greatest cost, —

---

* '88. Parsons, *Displacement and Area Curves of Fishes*, Trans. Am. Soc. Mech. Engineers.

enormous structural and physiological changes had of necessity to have been attained. The frame of the head and trunk has become moulded as in the fish's form, contours have been elaborately filled out and rounded, median dermal keels developed, vein valves lost, and the legs transformed into fin-like appendages.

The form of the fish is accordingly to be looked upon as cast in a more or less common mould by its environment. Its internal structures, as in the cetacean, are also observed to be modified in accordance with its external form. This is a factor in the evolution of fishes which appears in every group and sub-group. And it has ever stood in the way of classifying them satisfactorily according to their kinships.

"Fishes," used as a popular term, may include Lampreys, Sharks, Chimæroids, Lung-fishes, and "Modern Fishes" (Teleostomes), — the major groups to be discussed in the present book. But the relative position of each of these divisions must at present remain more or less doubtful. The group of the Lampreys is certainly widely removed from the remaining ones, standing midway between the simplest chordate, Amphioxus, and the true fishes : it is usually given a rank co-ordinate with either of these, and, in fact, with all other groups of vertebrates, taken collectively. Sharks, Chimæroids, Teleostomes, may be taken to represent true fishes ; and each might be assigned co-ordinate rank, although genetically the Chimæroids are certainly far more closely allied to the Sharks than are the Teleostomes. The Lung-fishes, as a widely divergent group, appear, as W. N. Parker has suggested, to be reasonably entitled to a rank equivalent to that of the three groups of true fishes taken together.

The present writer has, however, retained in the main the classification of Smith Woodward, in which Fishes (Pisces) is looked upon as a class, and is made to include as sub-classes, (I.) Sharks, (II.) Chimæroids, (III.) Lung-fishes, and (IV.) Teleostomes. A tabular grouping of the fishes is shown below. And on the opposite page their geological distribution is indicated.

TABLE I

A CLASSIFICATION OF FISHES

Type: **CHORDATA (VERTEBRATES).**

    Class: **Marsipobranchii, Lampreys,** *Palæospondylus*, Hag, Lamprey, *Ostracoderms*.

    Class: **Pisces (True Fishes).**

        I. Sub-class: ELASMOBRANCHII, Sharks and Rays.

            Order: *Pleuropterygii* (Dean), *Cladoselachids* (Dean).

              "   *Ichthyotomi* (Cope), *Pleuracanthids*.

              "   **Selachii,** Sharks and Rays.

       II. Sub-class: HOLOCEPHALI, Chimæroids, Spook-fishes.

            Order: **Chimæroidei,** *Squaloraiids*, *Myriacanthids*, Chimærids.

      III. Sub-class: DIPNOI, Lung-fishes.

            Order: **Sirenoidei,** *Dipterids*, *Phaneropleurids*, *Ctenodonts*, Lepidosirenids.

              "   **? Arthrodira,** *Coccosteids*, *Mylostomids*.

      IV. Sub-class: TELEOSTOMI, Ganoids and Bony Fishes (Teleosts).

            Order: **Crossopterygii,** *Holoptychiids*, *Osteolepids*, *Onychodonts*, *Cœlacanthids*.

              "   **Actinopterygii,**

              Sub-order: **Chondrostei** (Ganoids), *Palæoniscoids*, Sturgeons, Garpikes, Amioids.

              "   Teleocephali, recent Bony Fishes (Teleosts).

NOTE. — The groups italicized are represented only in fossil forms. The derivations of the scientific names are given on pp. 227–230.

TABLE II

THE DISTRIBUTION OF FISHES IN GEOLOGICAL TIME

The geological distribution of the prominent groups of fishes as here shown is in the main as given by Zittel (Palæontologie: Fische). The varying thickness of the lines denotes approximately increase or diminution in the number of existing genera.

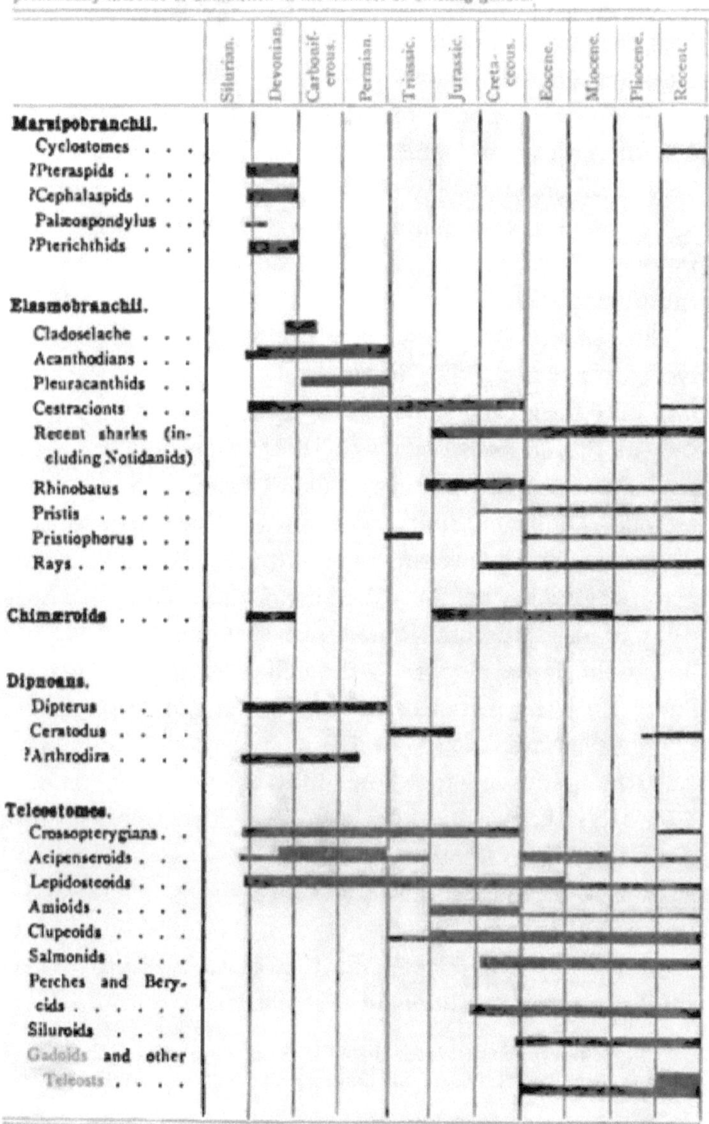

Fishes hold an important place in the history of back-boned animals : their group is the largest and most widely distributed : its fossil members are by far the earliest of known chordates ; and among its living representatives are forms which are believed to closely resemble the ancestral vertebrate.

The different groups of fishes appear especially favourable for comparative study. Their recent forms are generally well understood, both structurally and developmentally ; while a vast number of extinct fishes has been preserved to serve as a check, as well as an aid, to theoretical investigation.

The remarkable permanence of the different types of fishes seems a striking proof of how unchanging must have ever been the conditions of aquatic living. From as early as the Devonian times there have been living members of the four sub-classes of existing fishes, — Sharks, Chimæroids, Dipnoans, and Teleostomes. Even their ancient sub-groups (orders and sub-orders) usually present surviving members ; while, on the other hand, there is but a single group of any structural importance that has been evolved during the lapse of ages, — the sub-order of Bony Fishes. There are many instances in which even the very types of living fishes are known to be of remarkable antiquity : thus the genus of the Port Jackson Shark, *Cestracion* (Fig. 91), is known to have been represented early in the Mesozoic ; the Australian Lung-fish, *Ceratodus* (Fig. 127), dates back to Liassic times ;* the Frilled Shark, *Chlamydoselache* (Fig. 92), though not of a palæozoic genus, as formerly supposed (Cope), must at least be regarded as closely akin to the Sharks of the Silurian.

* Cf., however, Smith Woodward, *The Fossil Fishes of the Hawkesbury Series at Gosford.* Memoirs of the Geol. Surv. of N. S. W. Pal. No. 4, 1890.

The evolution of groups of fishes must, accordingly, have taken place during only the longest periods of time. Their aquatic life has evidently been unfavourable to deep-seated structural changes, or at least has not permitted these to be perpetuated. Recent fishes have diverged in but minor regards from their ancestors of the Coal Measures. Within the same duration of time, on the other hand, terrestrial vertebrates have not only arisen, but have been widely differentiated. Among land-living forms the amphibians, reptiles, birds, and mammals have **been** evolved, and **have** given rise to **more than** sixty orders.

The evolution **of** fishes has been confined to a noteworthy degree within rigid and unshifting bounds; **their** living medium, **with its** mechanical **effects upon fish-like** forms and structures, has for **ages** been almost constant in its conditions; **its changes** of temperature and density and currents have rarely been more than **of** local importance, and have influenced but little the survival of genera and species widely distributed; its changes, moreover, in **the** normal supply of food organisms, cannot be looked upon as noteworthy. Aquatic life has built **few** of the direct barriers to survival, within which the terrestrial forms appear to have been evolved by the keenest competition.

It is not, accordingly, remarkable that in their descent fishes are known to have retained their tribal features, and to have varied from each other only in details of structure. Their evolution is to be traced in diverging characters that prove rarely more than of family value; one form, as an example, may have become adapted for **an** active and predatory life, evolving stronger organs of progression, stouter armouring, and more trenchant teeth; another, closely akin in general structures, may have acquired more sluggish habits, **larger or** greatly diminished size, and degenerate characters **in its** dermal investiture, teeth, and organs **of sense** or progression. The flowering out of a series of **fish families** seems to have characterized every geological age, leaving **its clearest** imprint on the forms which were **then most** abundant. The variety that to-day maintains among the families of Bony Fishes **is** thus known to have been paralleled among the Carboniferous Sharks, the Mesozoic Chimæroids, and **the Palæozoic** Lung-fishes and **Teleostomes. Their environment has** retained their **general characters, while modelling** them anew **into** forms **armoured or scaleless, predatory or** defenceless, great, **small, heavy, stout, sluggish, light, slender, blunt,** tapering, **depressed.**

When members of any group of fishes became extinct, **those appear to have been the first to** perish which were **the possessors of** the greatest **number of** widely **modified or** *specialized* **structures.** Those, for example, whose teeth **were adapted for a particular kind of food, or whose motions were hampered by ponderous size or** weighty **armouring, were the first to perish in the** struggle for **existence ; on the other hand, the forms that most** nearly **retained the ancestral or tribal** characters — that is, those **whose structures were in every way least** extreme — were **naturally the best fitted to survive.** Thus *generalized* fishes **should be considered those of** medium size, medium **defences, medium powers of progression,** omnivorous feeding **habits, and wide distribution :** and these might be regarded **as having provided** the staples of survival in every **branch of descent.**

Aquatic living has not demanded wide divergence from **the ancestral stem, and the divergent forms** which may **culminate in a** profusion **of** families, genera, and species, do not appear to be again productive of more generalized groups. In all lines of descent specialized **forms** do not appear to regain by regression or degeneration the potential characters of their ancestral condition. **A generalized form** is like potter's clay, plastic in the hands of nature, readily to be converted into a **needed kind** of cup or vase ; **but** when thus specialized may **never resume** unaltered **its** ancestral condition : the clay survives; the **cup perishes.**

II

THE EVOLUTION OF STRUCTURES CHAR-
ACTERISTIC OF FISHES

It will be the object of the present chapter to review the gradations which occur in some of the characteristic structures of fishes and to follow in some degree the mode of their evolution. We may thus review the conditions of the (1) gills, (2) skin defences (including teeth), (3) fins, and (4) sense organs.

The structures of the immediate ancestor of the fishes cannot be definitely inferred: the form, however, must have been elongate and transversely jointed, for this condition seems to have existed remotely before fishes — in the broadest sense — had become evolved. This segmentation, or metamerism, of the vertebrate body is best shown among water-living forms, sometimes indeed in so perfect a way as to suggest the jointed condition of an earth-worm.

The segmented body of the eel-shaped Lamprey, shown in section in Fig. 69, illustrates an interesting condition of vertebrate metamerism. Its entire body, from the head region to the base of the tail, is composed of drum-like segments which closely correspond to one another in size and in component structures. Each segment thus resembles its neighbours in its equal portions of the vertebral column, digestive tract, nerve tube, muscle plates and blood canal, and in the arrangement of these parts with reference to bilateral symmetry. Motion in this form requires no more of each segment than that its sides contract alternately **to** produce **a** rhythmical **wave** passing along the entire series of segments and giving the trunk an undulatory movement.

Should this elongate body now **acquire** a more fish-like **form,** in attaining, for example, **the power of more rapid** movement, it is obvious that this simple type of meta-merism would undergo a series **of** changes. Every change of outward form would be reflected on the parts not only of each, but of all segments in their common relationships. To perform more perfectly the functions of their location, adjacent segments might become enlarged, **folded, or** blended, **and cause the** most puzzling complications of their component structures. **One region of the** body might thus appear to develop at the **expense of another,** as in the evolution of fin structures (cf. **pp. 32–44), where a vertical** fin fold, representing the sum of the dorsal and ventral out-growths of the hinder body segments, becomes reduced to the lappet-like dorsal and ventral fins; the intervening substance of the fin web becoming drawn to the points where greater rigidity **is required.**

The simple metameral character of the lamprey acquires an especial interest when the different groups of **fishes are** examined; for it **is** found that all exhibit clearly **body** segments and segmental structures in the most **varied** stages of complexity. To trace metamerism seems, accord-ingly, a mode of determining to what degree the differ-ent groups have diverged from a common stem; and to compare the sums of the archaic metameral characters in **the** different types of fishes **may** perhaps be looked upon as one of the safest aids **in** determining **their** genetic posi-tion. From the conditions of segmentation the lampreys must certainly be given **a lowly rank; even with** due allow-ance for degeneration **of** structures **they are clearly more** primitive than the most archaic sharks: while, on the other hand, to the metameral type **of** the sharks may the structures of the remaining groups of fishes be best referred.

1. AQUATIC BREATHING

Respiration in fishes is developed on the primitive chordate plan **of** ejecting water through gill slits perforating the throat wall. The water taken in by the mouth is rich in absorbed air, and, as it passes out, **is** well calculated to **oxygenate the blood suffusing** the sides **of** the gill slits.

Among the earliest chordates **there seems** evidence **that the** gill openings **of the** gullet **were** arranged **with reference to some** form of primitive segmentation. Perhaps they occurred as **well in the** region **of the** mid-digestive **tract, before their location** became restricted to the **gullet. There has been as** yet, however, little satisfactory evidence * **as to** the number or conditions of the gill slits in very **primitive** forms. In Amphioxus the gill arrangement **seems clearly** a most specialized **one**: its adult condition **presents an atrium and an elaborate** branchial **basket,†** which could **hardly have occurred in** the **lowly** ancestral chordate. **Its early larva, however, is** known **to** possess (but in a condition **of assymmetry)** but a few gill slits (seven to nine) from which the many openings of the adult branchial basket take their origin, — a developmental **stage** which most closely and most interestingly suggests the conditions of higher forms.

* **It has generally** been inferred that the immediate ancestors of fishes had **not many gill slits,** probably not more than eight or nine. A Liassic shark, a **Cestraciont,** *Hybodus* (p. 85), is known to have had but five; a Permian *Pleuracanthid,* **as in** the recent *Heptanchus,* seven **(p. 88)**; the Lower Carboniferous *Cladoselache* probably seven.

† *Cf.* Vol. II, of this series. Willey, *Amphioxus and Other Ancestors of the Chordates.*

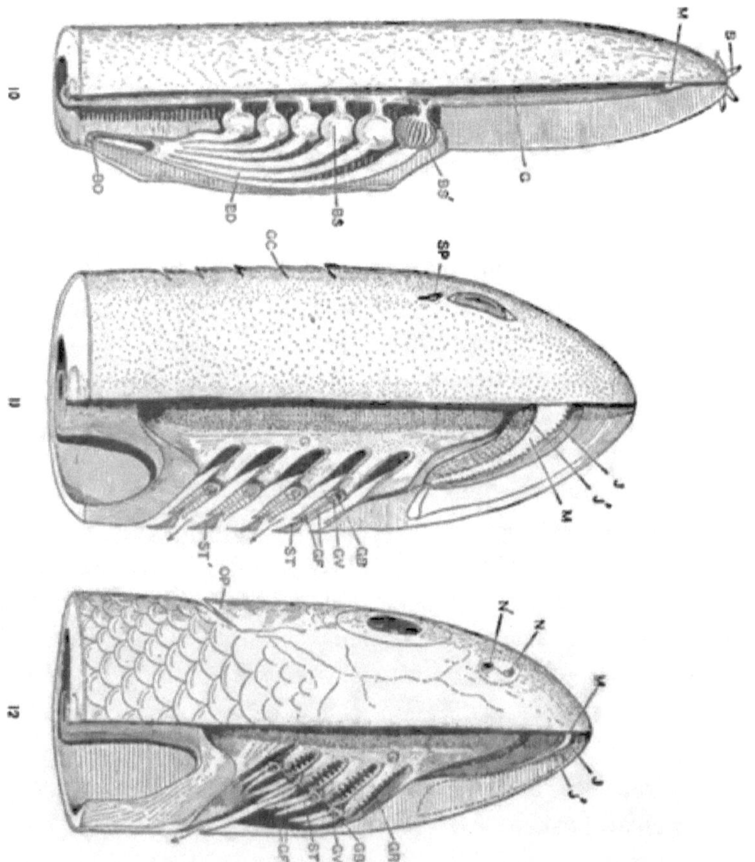

Figs. 9-12.—Arrangement of gills of Bdellostoma (9), Myxine (10), Shark (11), and Teleost (12). In each figure the surface of the head region is shown at the left.

B. Barbels. *BD.* Outer duct from gill chamber, *BS. BO.* Common opening of outer ducts from gill chambers. *BS.* Branchial sac, or gill chamber. *BS'.* Branchial sac, sectioned so as to show the folds of its lining membrane. *G.* Lining membrane of gullet. *GB.* Gill bar, supporting vessels and filaments of gills. *GC.* Outer opening of gill cleft. *GF.* Gill filament. *GR.* Gill rakers. *GV.* Vessels of gill. *J, J'.* Upper and lower jaw. *M.* Mouth opening. *N, N'.* Anterior and posterior opening of nasal chamber. *OP.* Operculum. *SP.* Spiracle. *ST.* Tendinous septum between anterior and posterior gill filaments. * Denotes the inner branchial opening; →, the direction of the water current.

In the singular group of lampreys and slime eels (Marsipobranchs, v. p. 57), the segmental arrangement of the gills seems of a primitive pattern. In the Californian Myxinoid (p. 59) the slits are as numerous as thirteen and fourteen on either side, each opening directly from the gullet to the neck surface (Fig. 9, *G*, *, *BS'*, *BD*). In the lamprey the conditions are similar, but the number of gill slits is reduced to seven. In Myxine (Fig. 10, *G*, *BS'*, *BD*, *BO*) the outer portions of the canals becoming produced tail-ward have merged in a single pore (Fig. 71 *). In these forms each gill canal has become dilated at one point of its course, and in this sac-like portion the blood-suffused tissues have grouped themselves into leaf-like plates (gill filaments, or lamellæ, *BS'*) to increase their surface of contact with the out-passing water. The dilating power of this gill sac has then become specialized so that even should the animal's mouth be closed, water for respiration could be drawn in through the canal's outer opening: from this acquired function the elaboration of branchial muscles and a supporting framework of cartilage (branchial basket, Fig. 69 *A*, *BB*) may have taken its origin.

Among fishes proper many stages in the evolution of gill organs are represented. They show altogether a marked advance over the conditions of Fig. 9. There has been a general tendency to press closely together the gill pouches and to elaborate into thinner and larger lamellæ the blood-suffused tissue. In this process the gill chamber has become slit-like, bearing gill lamellæ only on its front and rear margins; its supporting tissue has consolidated into stout vertical gill bars, the gill structures in general, becoming more highly perfected, tending to recede from the surface. These conditions may best be illustrated by contrasting the highly modified gill apparatus of a bony fish with the more archaic type of the shark.

In the sharks (p. 73) the gill slits pierce separately the throat wall, as in the lamprey, and thus retain their primitive segmental arrangement (Fig. 11). Their number is usually five on either side, but in an archaic form (Heptanchus, p. 88) may be increased to seven. Above and in front of the line of gill slits occurs a small opening leading into the gullet, the *spiracle* (*SP*). This, though at present possessing but few gill lamellæ, and therefore of little respiratory value, was doubtless quite like its neighbours before its gill-supporting tissue became of value in suspending the lower jaw. It may now aid the mouth opening in admitting water to the gills. At the left of the figure (Fig. 11), the narrow slit-like openings of the gill clefts are seen at *GC*: at the right, where the upper portion of the head has been removed, the gill lamellæ are shown at *GF*; the tissue intervening between the gill pouches is reduced to a thin tendinous septum, *ST*, at whose inner rim is the cartilaginous gill arch or bar, *GB*, supporting the branchial vessels, *GV*.

In the gill region of a bony fish (Fig. 12) a number of modified characters are now evident: the spiracle has become obliterated; the number of gill bars reduced — in one form but two on either side remaining. These have become closely pressed together, and bent backward, receding from the surface of the head: their gill lamellæ have become larger and more numerous, their intervening septum, *ST*, reduced in size. The gills no longer open separately at the surface, but into an outer branchial chamber formed and protected by a large overlapping scale, or opercle, *OP*. This shield-like organ is hinged at its anterior margin and opens or shuts rhythmically as the throat muscles draw in or eject the water used in respiration. On the gullet wall, the gill bars, now seen to be closely drawn together, have acquired marginal outgrowths, or gill rakers, *GR*, which form an interlocking screen across the gill openings and prevent the escape of food organisms. So perfect may this apparatus become that the opening and closing gill bars may retain even microscopic life.[*]

Between the conditions of Figs. 11 and 12 there occur many transitional forms.

To protect the gill region, specialized devices are known to have been evolved early in the history of fishes, — the more early if, as Garman has supposed, the gill filaments in primitive sharks protruded at the sides of the head.[†] There are thus the gill-encasing derm frills of the archaic sharks, *Cladoselache*, *Chlamydoselache*, and *Acanthodes* (pp. 78–83), or of Chimæroids (p. 100). These protective structures, the writer believes, may well have originated independently even within the limits of subgroups. They have certainly no direct relation to the opercle of bony fishes.

Modes of respiration by gill filaments have been found in endless variety among fishes, clearly dependent in the majority of cases upon environment. Thus fishes that require a temporary existence out of water will be found to have specialized spongy gill filaments and a closely fitting gill cover to keep moistened the respiratory organs (*e.g. Callichthys*, p. 172).

---

[*] Thus in many bony fishes, *e.g.* mullet or Brevoortia (menhaden), the inner margins of the gill bars are fringed with what appears like the finest gauze, each gill raker giving off primary, secondary, and tertiary branches. A somewhat similar condition occurs in the shark, Selache (p. 90).

[†] This condition appears to have been possessed by the Lower Carboniferous Cladoselache.

To live a longer time out **of** water has been rendered possible only by the appearance of a lung-like organ.   Such a structure, however, would have been of too great importance in **the** living economy of terrestrial vertebrates to **have** had a sudden origin : **it may** most reasonably have been derived from a similar structure occurring very generally among fishes.   The lungs certainly resemble the swim-bladder of **fishes** in so many important characters that it seems difficult to regard these organs as morphologically distinct.   **In** itself the swim-bladder must be looked upon **as an ancient** and essentially a **generalized structure,** for within **the groups** of fishes it has already **acquired a variety** of modified characters : appearing **in a lowly condition** in sharks, **it** acquires a balancing function **in the majority of bony fishes ;** in some **forms (carp, siluroids) its function** connects it with the auditory **organ, often by a** highly elaborated apparatus : while **in** other forms (*Amia*, Garpike, Dipnoans), it is unquestionably of respiratory value. **The** wide range in the characters of the air-bladder (cf. Figs. **13–19,** and Table, p. 264), even among recent fishes, would naturally favour its homology with the lungs : it may thus be paired or unpaired, attached by its **duct to either** the dorsal, lateral, or ventral wall of the gullet : it may present the most varied characters in its lining membrane or in its vascular supply.   When, moreover, it becomes of respiratory value (*e.g.* Dipnoans, *Polypterus*), the gills are **known to** become in part degenerate.   The larval history of amphibians, presenting so perfect **a** transition between gill-breathing and terrestrial vertebrates, **should alone seem to** render more than probable the general homology of airbladder and lung — an homology which a closer knowledge of the conditions of the lungs of the lower urodeles (*e.g. Necturus* may well **be** expected to establish definitely.

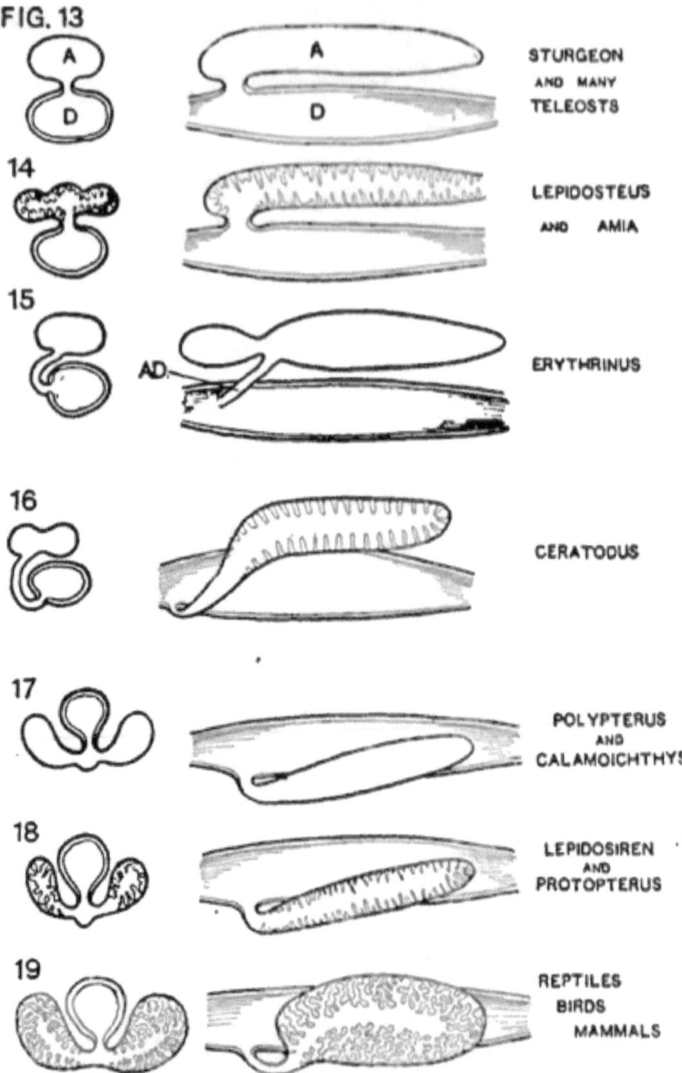

Figs. 13-19.—Air-bladder of fishes, shown from the front and sides. Cf. p. 264. *A.* Air- or swim-bladder. *AD.* Air duct. *D.* Digestive tube. (After WILDER.) 13. Sturgeon and many Teleosts. 14. Amia and Lepidosteus. 15. Erythrinus, a Cyprinoid Teleost. 16. Ceratodus. 17. Polypterus and Calamoichthys. 18. Lepidosiren and Protopterus. 19. Reptiles, birds, and mammals. The diagrams illustrate the paired or unpaired character of the organ, its varied mode of attachment to the digestive tube, and the smooth or convoluted condition of its lining membrane.

The mode of origin of the lungs as an unpaired diverticulum of the gullet is in every sense similar to that of the air-bladder.

2. THE DERMAL DEFENCES OF FISHES

The dermal defences of fishes include scales, spines, fin rays, armour plates, and teeth, presenting in all a wide range of calcified structures. They have usually an outer, or surface layer of hard enamel-like texture and an inner substance heavy, stout, and bone-like. The former is derived from the outer layer of the skin (epidermis), the latter from the derma. The relation of these structural parts may be well seen in a section of shark skin which passes through one of its minute limy cusps, or dermal denticles (Fig. 20). The outer skin layer, E', originally covered the denticle, which grew outward, papilla-like, beneath it ; its inner surface, in contact with the outgrowing papilla, secreted the enamel, E, and is known as the enamel organ, EO: at the cusp, however, the epidermis is early worn away. The bone-like substance of the tooth is clearly formed in the lower (dermal) layer of the skin, D': it is formed by the calcification of the outer layers of the tip and base of the dermal papilla, leaving a vascular cavity, PC, within. This limy substance, "dentine," D, presents microscopically a columnar "cancellated" structure; in this and in its lack of bone cells it differs structurally from true (cartilage) bone.

The dermal denticle of the shark is certainly the simplest form of a calcified skin defence : it appears to represent the ancestral condition of the various scales, teeth, or bone plates which have been evolved in the groups of fishes. It is usually of minute size, and studs closely the entire surface of the skin, forming shagreen. In many

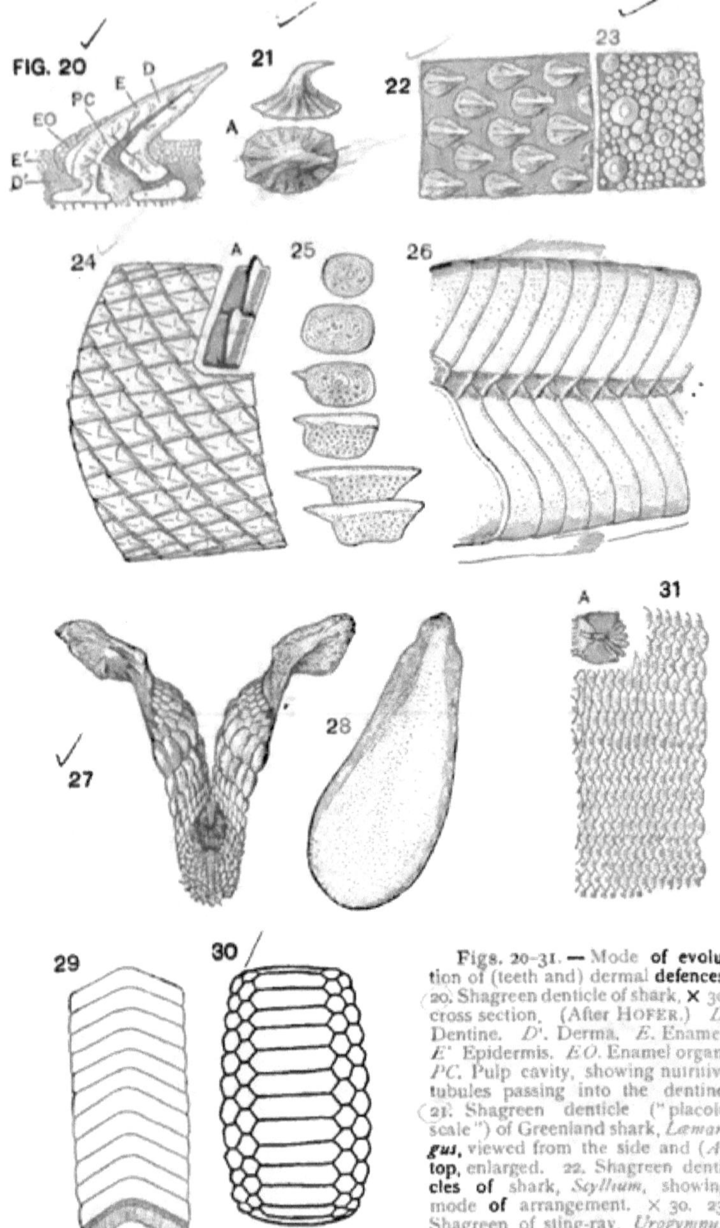

Figs. 20–31. — Mode of evolution of (teeth and) dermal defences. 20. Shagreen denticle of shark, × 30, cross section. (After HOFER.) *D.* Dentine. *D'.* Derma. *E.* Enamel. *E'* Epidermis. *EO.* Enamel organ. *PC.* Pulp cavity, showing nutritive tubules passing into the dentine. 21. Shagreen denticle ("placoid scale") of Greenland shark, *Læmargus,* viewed from the side and (*A*), top, enlarged. 22. Shagreen denticles of shark, *Scyllium,* showing mode of arrangement. × 30. 23. Shagreen of sting-ray, *Urogymnus,* nat. size. (After SMITH WOODWARD.) 24. Ganoid dermal plates of *Lepidosteus.* *A.* Inner face of ganoid plates, showing tile-like device of interlocking. 25. Variation of ganoid plates in *Aetheolepis.* (After SMITH WOODWARD.) Plates from different regions vary in outline from circular to lozenge shape. 26. Coalesced ganoid plates of the siluroid *Callichthys.* 27. Jaw of Port Jackson shark, *Cestracion.* 28. Dental plate of extinct cestraciont (?), *Sandalodus.* 29. Dental plates of jaw of sting-ray, *Trygon* (?). 30. Dental plates of eagle-ray, *Myliobatis.* 31. Scales of Teleost. *A.* A single scale enlarged.

members of the shark group the denticles are scattered over the body without traces of metameral arrangement (Fig. 23); in others they acquire a segmental position (Fig. 22). Usually the denticles possess very definite shapes and regional characters; their basal portion, where implanted in the skin, may thus become of enlarged size and regular outline (Fig. 21 *A*), their projecting cusps tapering, blunted (Fig. 23), or branched. Sometimes the fusion of contiguous denticles may occur (as in the enlarged blunted denticles of Fig. 23).

The evolution of the more perfect body armouring of fishes from shagreen denticles has not been followed in minor details. It appears, however, that the calcification of the skin which occurs superficially in the dermal papillæ of the shark may in other fishes be traced occurring in deeper and deeper layers of the derma: the papillæ at the surface accordingly lose their functional importance, and tend to disappear, while the calcified tissue of the derma — representing morphologically the basal region of the denticles — is coming to occupy more and more definite tracts. These processes have already taken their origin within the group of sharks.

An interesting condition in the subsequent evolution of the dermal armouring is illustrated in Fig. 25, and has been described by Smith Woodward. The circular bone plate of the figure is a calcified dermal tract which still retains, scattered generally over its surface, traces of shagreen tubercles: from this shark-like condition a well-marked gradation in the form of the derm plates may be traced in different body regions of the same fish: according to metameral needs there are acquired rectangular or lozenge-shaped outlines. In Fig. 24 these bone or "ganoid" plates are seen to constitute a complete but flexible body armouring, made additionally strong by an interlocking articulation of its elements (24 *A*).

In this form the enamel-like surface layer ("ganoine") of the ganoid plates is believed to be derived from the dentine substance, and not deposited by the epidermis: they bear numerous shagreen denticles during an early period of life.

The most complete encasement of a fish's body by dermal plates is shown in Fig. 26, v. p. 172. The met-ameral conditions have here permitted extended fusions, a single dermal plate enclosing the upper, or lower division of the muscle-plate of either side.

The thin horn-like scales of the majority of recent fishes, *e.g.* carp or perch (Fig. 31 *A*) are probably derived from a condition not widely different from that of Fig. 24. They take their origin, however, in a deeper layer of the derma, thence grow outward, arising as if from deep and flattened pockets. Their substance becomes horn-like, rather than limy, and they enlarge in outline, rather than in thickness. Their hinder margins, often crenulate, overlap widely the neighbouring scales; their arrangement is in direct relation to the underlying metameres, and their surface is densely slime-coated. The dermal armouring they thus constitute is both light, tough, and flexible.

Degeneration of scales is shown to occur in many types. In some forms their size may become micro-scopic (eel), in others enormously enlarged (mirror carp). In cases they may entirely disappear (leather carp). The fusions of the dermal plates of the trunk-fish or of the sea-horse (p. 177) are probably degenerate.

Teeth

Teeth have long been known **to represent the dermal** defences of the mouth rim. **In** this region they have **become of** especial value **in** the living economy **of vertebrates** — seizing, holding, cutting, or crushing the food-material. They have here accordingly been retained and specialized. **In** the sharks the dermal denticles of the mouth rim are often identical in shape and pattern with those of the entire body surface: they differ only **in** their larger size. Their arrangement **in** many rows still presents clearly their metameral character.

The forms of teeth acquired among the different groups of fishes suggest closely the evolution **of the more modified** dermal defences. In general, they are **found to vary** widely according to their function **or** location; **those nearest** the dermal margin of the mouth usually retaining **the** cusp-like and more primitive features. Thus in the **jaw of** Port Jackson **shark** (Fig. 27, v. p. 85), the teeth of **the** symphysial region clearly represent shagreen denticles; while those deeper in the mouth, large and blunt, serve as crushing or "pavement" teeth. These must evidently be looked upon as standing in the same relation to the anterior cusps, as do the bone plates of Fig. 25 to the derm denticles of Fig. 23; the fused crushing teeth have still retained their metameral arrangement. The dental plates (Fig. 30) of a ray, *Myliobatis* (p. 96) show **more** perfect conditions for crushing; they are uniform **in size,** tightly set, and present a smooth, mosaic-like surface. **A** still more perfect **fusion of the** dental elements **occurs in** a ray, closely akin **to Myliobatis;** all lateral elements **have** here been fused, but their metameral sequence **has been retained** (Fig. 29). In Fig. 28 is **shown a** dental plate of **a**

fossil shark (?), *Sandalodus*, which probably represents a condition of complete fusion; it would accordingly correspond to the sum of the dental elements of half of the jaw of Fig. 27.

In more highly modified fishes the tooth-producing region has become greatly extended; teeth are present not only on the jaw rims, but deep in the mouth cavity, studding its floor and roof, and occurring even on the tongue, gill bars, and pharynx.

Fin Spines

Primitive dermal defences appear to have played a prominent part in the formation of fin spines. The clustering of dermal cusps on the exposed margin of a fin may have been an important initial step toward the formation of a rigid cutwater. The anterior margin of the fin of Fig. 49 is whitened with a fusion of dermal tubercles which must have formed a firm encrusting support; the extension of the calcification of the bases of the tubercles would accordingly be the mode of origin of a fin spine. In Fig. 32 is shown a spine that appears largely of this origin. A similar spine (Fig. 33) shows its dermal tubercles not only at its sides, but in a most marked way at its hinder margins. In Fig. 34, representing the "sting" of the sting ray, a series of dermal spines, bearing rows of minute denticles are seen to arise in a metameral succession. A condition somewhat similar is known in the Carboniferous shark, *Edestus* (Fig. 35), whose spine, often of gigantic size, is of special interest, since it shows how important a part in spine-formation may be taken by the dermal defences of many successive metameres. The spine is clearly segmented, and as its separate elements (Fig. 37) are bilaterally symmetrical (Figs. 36 and 38), its position was probably in the median line of the body. The well-marked, backward curve of the spine suggests

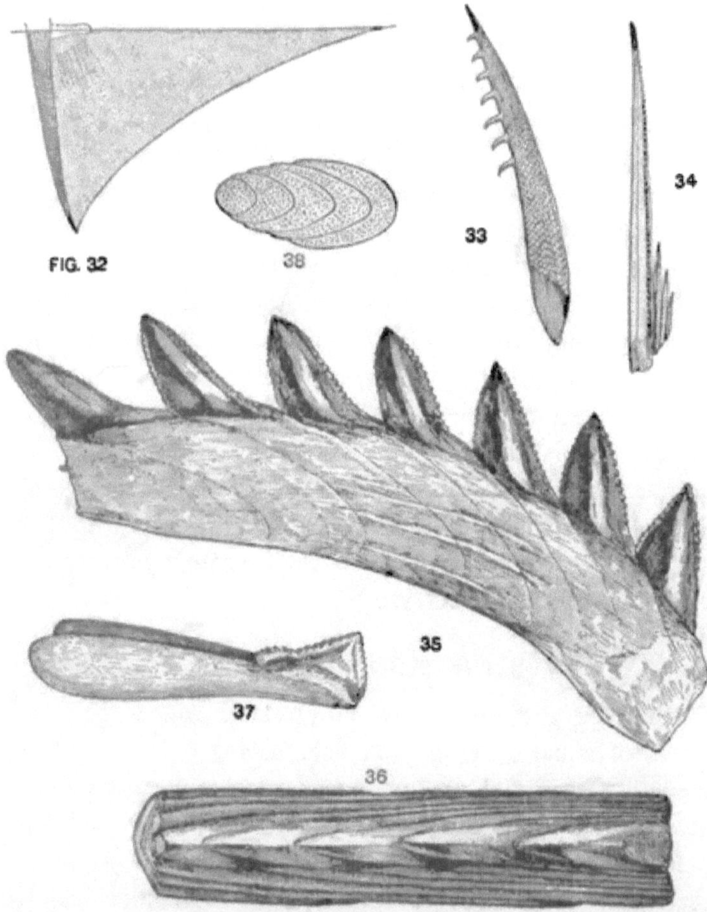

Figs. 32–38. — Fin spines. 32. Fin spine and pectoral fin of Acanthodian. 33. *Hybodus* (cestraciont shark). 34. Sting-ray, *Trygon*. 35. *Edestus heinrichsii* (Carboniferous shark, known only from its spine), side view of spine. × ¼. 36, 37, 38. Dorsal view, separated element and transverse section of Edestus spine.

that fin structures could not well have existed behind it. Each separate element has an elongated basal portion, which apparently was imbedded in the integument; its gouge-like form (Figs. 37 and 38) permitted it to be firmly apposed to its anterior and posterior neighbours. Each median enamelled cusp represents apparently the sum of the shagreen papillæ, occurring in the median-dorsal region of each metamere, its gouge-like underlying portion the metameral calcification of the bases of the denticles.

What has been the mode of origin of the primitive derm cusps is a puzzling question. It is significant, perhaps, that they occur in primitive forms (sharks) in connection with the sense organs of the lateral line (p. 50), and that they are in this region retained in a number of archaic forms (Polypterus, p. 148, Callichthys, p. 172), which have in all other body parts evolved protective derm plates.* It is certain that for the sensory groove of the lateral line, no more simple, protective devices could have arisen than conical elevations of skin. Arising in this region, they may have extended their protective functions over the entire body surface.

3. THE EVOLUTION OF FINS

Fins are the organs of progression adapted to the needs of aquatic living. A fish, balanced in its living medium, acquires, as has been seen, a boat-like form, enabling it to pierce the water in the least resisting manner. Its appendages, when they come to arise, must reasonably be looked to to fulfil the mechanical conditions of aquatic motion in order to propel to the best advantage the lightly balanced and boat-shaped mass. Fins might thus be expected to arise as keel-like struct-

---

* In the sensory canals of the head of Chimæra, the presence of scattered bony plates, protective in function, v. p. 114, would suggest the concentration of the marginal cusp elements for more perfect protection.

ures, *i.e.* as ridges in **the** direction **of the** fish's axis **or** line of motion.

Fish fins have long been distinguished as vertical (median, or unpaired) or lateral (paired), the former functioning both as keel and means **of** propulsion, the latter **as** accessory and specialized balancing organs.

Median Fins

Median fins are unquestionably the older. They exist **in** the simplest condition in those fishes whose axis is long and whose motion is undulating. **Indeed, the sole swimming** requisite **is** here the continuous **dermal keel which** passes **down** the **back from the head to the body terminal,** and **extends** thence forward on **the ventral side. The** undulatory motion of the body is well **transmitted to the** surrounding medium by **the** exaggerated **undulation of** this long, waving fin web. This condition was probably the ancestral one in the evolution **of** fishes. It represents the simplest metamerism ; it occurs as the adult condition in the lampreys (p. **57),** and as the embryonic or larval stage **in all** fishes, appearing **before any** traces **of paired** fins are known ; it is **even adverse to their** specialization : should life habits require undulatory **motion,** paired fins must inevitably tend to disappear **(eel, p. 173 ;** Calamoichthys, p. 150).

From this condition the further evolution of the **un**paired fins may thus be theoretically outlined.

The primitive continuous dermal **fin** could have **been** of little value in active movement its **more** rapid **undu**lations could not have greatly increased the rate **of** motion, since its web, lacking in supports, would not have retained its rigidity. **As the** simplest means of strengthening the fin fold, *"actinotrichia"* (Ryder), appear to have been early evolved (Fig. 39, *T*) ; these are slender, unjointed fin supports, passing from the body wall to the margin of the fin, appearing to arise without relation to the underlying body segments. The more rapid undulations of the continuous fin would next cause nodes to arise ; and at other points the greatest mechanical stress would occur. These portions of the fin web would accordingly become prominent, while the intervening or useless parts would diminish in width and tend to disappear. The body terminal (tail, caudal fin) has now become the seat of propulsion: dorsal and ventral fins arise as lobate elements of the fin fold, functioning as vertical keels in the region of the body where mechanical stress demands them (v. Fig. 40), increasing in size as the intervening portions of the web gradually disappear. Their rate of growth is doubtless affected by the appearance of the paired fins ; for even at an early period of development these are known to have an important function in balancing the fish.

The lappet-shaped fins (Fig. 40) next acquire more rigid supports. Cartilaginous rod-like elements arise within the fin web, arranged in metameral sequence, representing, perhaps, fusions of actinotrichia. As shown in Fig. 40, these cartilaginous "*radials*," *R*, appear to be largest and stoutest in the widest por-

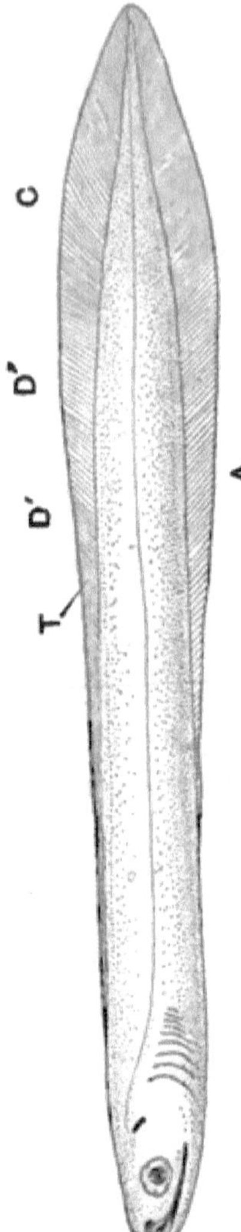

Fig. 39. — Hypothetical ancestral shark. Letters as on p. 33.

tions of the fin lobe, and thence to taper in size toward the nodal points of the web. Each radial appears shortly to segment off a proximal joint, or "*basal*" cartilage, *B*, to secure a more perfect attachment with the wall of the body.

The subsequent evolution of the fins appears to have been determined by two modifications of growth, — the clustering of the radial and basal elements, and the encroachment of newly formed marginal (distal) rays

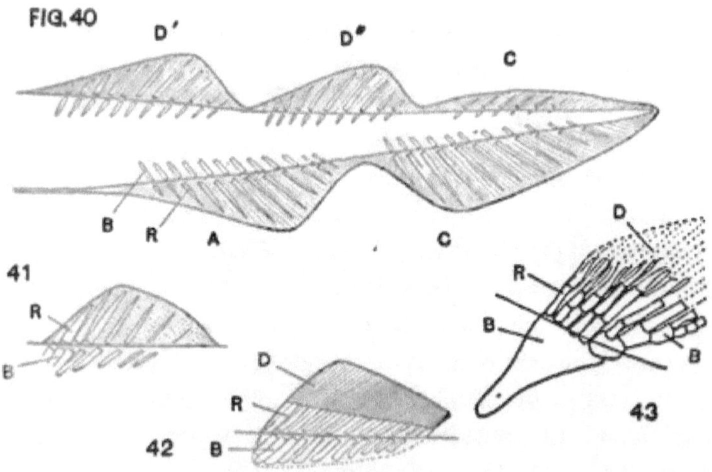

Figs. 40-43. — Evolution of unpaired fins. 40. Plan of reduction of vertical fin web into its dorsal, anal, and caudal elements. 41. Arrangement of fin supports in primitive fin (*Cladoselache*). 42. Plan of archaic unpaired fin in (larval) shark. 43. Unpaired fin of fossil Crossopterygian, *Holoptychius*. (After SMITH WOOD-WARD.)

A. Anal fin. *B*. Cartilaginous basal (fin support). *C*. Caudal fin. *D*. Dermal margin of fin. *D'*. Anterior and *D"*. Posterior dorsal fin. *R*. Cartilaginous radial (fin support). *T*. Actinotrichia.

upon the functions of the older fin supports. Three stages in this metamorphosis will be seen in Figs. 41-43. The first illustrates the dorsal fin of an ancient shark (Cladoselache, p. 79), and will at once be seen to present most primitive conditions : it closely resembles the theoretical dorsal fin, D' or D'' of Fig. 40. The form of the fin suggests the lobate constriction of the continuous fin web; its radial supports, R, extend from the body wall to the margin of the fin, and between them traces of actino-trichia are to be seen. The anterior margin of the fin must now function as a strong cutwater, its supporting elements, both radial and basal, tightly clustering. A fin of this character could evidently have possessed a greater freedom of lateral movement in its hinder than in its anterior part; and thus the clustering of the fin supports becomes of especial significance. The region of movement, restricting itself to the hinder part of the fin, permits extensive fusions of the supporting cartilages anteriorly, and leads ultimately to exceedingly complex conditions. The dorsal fin of a Coal Measures fish (Holoptychius, p. 151) has thus (Fig. 43) specialized the power of lateral movement in the highest degree. The length of the fin has, in the first place, become greatly compressed, a process which seems to have resulted in implanting the anterior basals, B, deeply into the integument and in fusing them: the posterior basals then appear to have been everted from the surface of the body. Here they still retain their segmental arrangement, but are irregular in shape and reduce in size distally.

An important part is taken by the *dermal* margin of the fin in modifying the size of the older fin supports. The simplest form of a dorsal fin of a recent shark (Fig. 42) has thus more than half of its functional area of a dermal origin, although. in other regards it resembles closely the conditions of Fig. 41. The dermal margin of the fin has apparently increased to the detriment and consequent reduction of the cartilaginous elements; it produces in its secondary structures light flexible horn-like rays, which prove stronger and more serviceable **than** the heavier radials; it seems more capable **of** adapting the fin for special uses.

Accordingly, in many forms of recent fishes, notably bony fishes, the entire fin **is found to become of** dermal origin; the radio-basals, greatly reduced in number and size, extend no further outward than the base of the fin; they are usually small **and** irregular, and are often deeply sunken within the body wall.

After this glimpse at the mode of origin of the vertical fins, *i.e.* dorsals and anals, the history **of the final** vertical fin, the tail, and of the paired **fins may next** be reviewed.

The Caudal Fin

The tail, or **caudal fin, is the main organ of** aquatic propulsion, and it **is doubtless on this account that it** presents so wide **a range in its structure and outward** form. From the earliest times **there** are found fishes of all groups whose tail shapes **are** tapering (*diphycercal*, Fig. 47), unsymmetrical (*heterocercal*, Figs. 45, 46), or squarely truncate (*homocercal*, Fig. 48), as the mechanical needs in swimming may **have** demanded.

The following summary of **the mode of** evolution of the caudal fin seems to be warranted by study of fossil and embryonic **forms.** The vertical fin fold of the ancestral fish was probably carried around the body terminal and strengthened by constant actinotrichia (Fig. 39 *C*), a condition similar to that (Fig. 44) of an early larval stage of living fishes (*protocercy*). This caudal structure, however, could have proven of value only in sluggish undulatory motion. The functional needs, **which** gave rise to radials anteriorly, have in the tail region produced firmer and stouter fin supports. These appear both on the dorsal and ventral sides, but, unlike the radials of the **anal** or dorsal fins, do **not** segment off basal elements. They first occur **in the** region **of the** base of the caudal, as in the embryonic stage (Fig. **44,** *R*), since, perhaps, it is in this **region** that the greatest stress occurs in propulsion. **It is** not until a later stage that **their** metameral sequence **is extended** backward **to the tip of the** vertebral axis (Fig. **40,** *C*).

With the origin of cartilaginous supports there seems to have arisen a mechanical need for enlarging the ventral lobe of the caudal; it is here certainly that in the majority of early forms the radials appear longer **and** stouter, giving rise to the condition **of heterocercy** of Figs. 45 and 46. The greater functional importance of the radials of the ventral region, $R + H$, is acquired contemporaneously with the upturning of the end of the vertebral **axis.** In the tail of a **Lower** Carboniferous shark (Fig. 46, **v.** p. 79), **an** extreme degree of heterocercy **has been** acquired before the radials of the lower lobe have extended themselves in the hindmost region of the vertebral axis; the ventral web of the upper tail lobe, **accordingly, is** still strengthened by minute (dermal) rays, which the writer believes homologous with actinotrichia; on the fin's dorsal side the radials have been abruptly upturned with the notochord, and are fused into a compact cutwater.

The plan of structure of the shark's caudal fin (Fig. **45)** may in its most primitive form prove to be the ancestral one of fishes; if this is the case it would give rise to the types of caudal fins of Figs. 47 and 48. That it has given rise to the latter form cannot be doubted, for even in the adult condition of the fin the notochord, *N*, may be seen passing to the upper lobe of the tail; the essential outward form **of** this truncated, or homocercal, tail had already

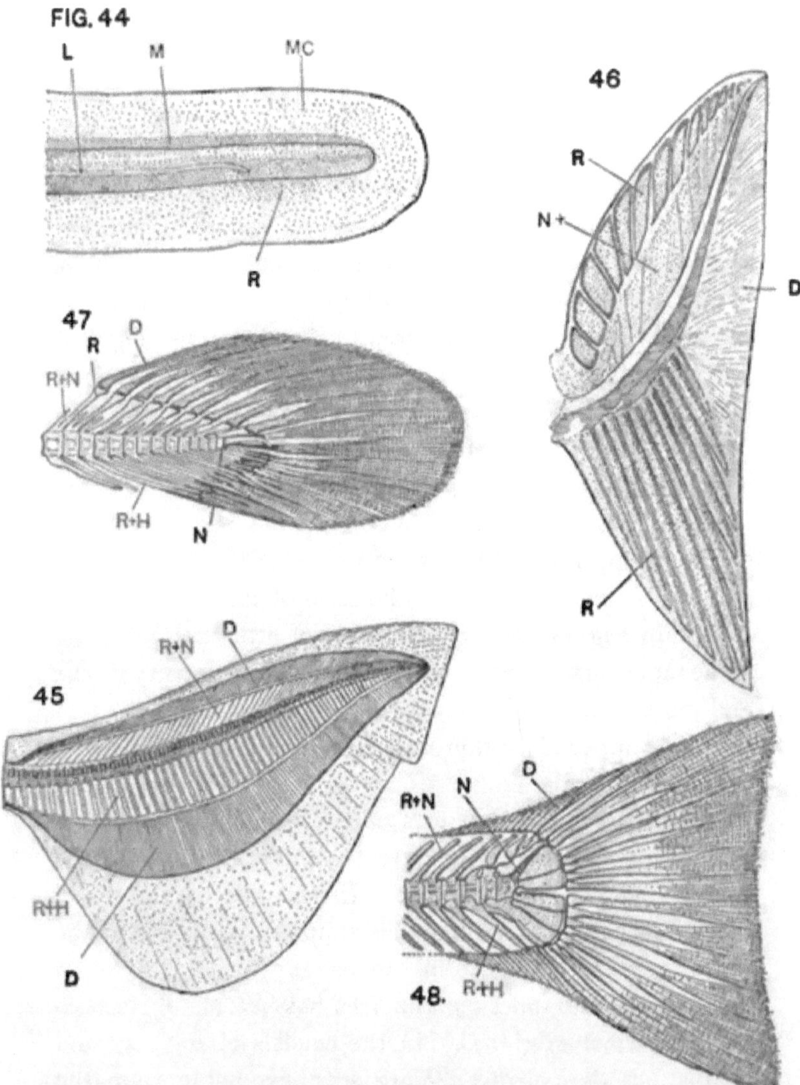

Figs. 44-48. — Evolution of caudal fin. 44. Embryonic caudal of *Amia*. 45. Heterocercal caudal of shark, *Cestracion*. 46. Heterocercal caudal of *Cladoselache*. 47. Diphycercal caudal of Polypterus. (After L. AGASSIZ.) 48. Homocercal caudal of Teleost. (After RYDER.)

D. Dermal fin supports. *L*. Lateral line. *M*. Spinal cord. *MC*. Membranous caudal. *N*. Notochord. *N+*, and *R+N*. Neural spines, including probably radial and basal elements. *R*. Radials. *R+H*. Hæmal arch and spine; includes as well, probably, radial and basal elements.

been acquired in ancient sharks (Fig. 46). The fin of Fig. 47, however, has not generally been looked upon as derived from shark-like conditions ; it has, on the other hand, been thought to be most nearly of the ancestral form. The vertebral axis does not appear to be upturned, and the ventral and dorsal lobes of the fin remain nearly symmetrical, or diphycercal. This form of the caudal fin, on the other hand, has been noted to present many degenerate characters, and to the writer * it seems more reasonable to regard the diphycercal condition as in many cases directly descended from the heterocercal. This might be effected by the terminal portion of the vertebral rod aborting (as in Fig. 47, *N*), and the upper and lower lobes of the tail becoming pressed backward until their hinder margins appose in the axial line.† The form of diphycercy which is seen in Fig. 119 is unquestionably of little morphological value ; it occurs commonly in deep-sea fishes of every group, and must be looked upon as a degenerate condition resulting from impeded motion under the conditions of bathybial, or deep-sea living.

The cartilaginous supports of the caudal, like those of other unpaired fins, become greatly reduced in size by the encroachment of dermal rays. In the tail of the fossil shark (Fig. 46) the cartilaginous supports, *R*, extend to the very margin of the fin : in the modern shark (Fig. 45) a large part of the functional fin area has become of secondary, or dermal origin, *D*. In the caudals of Figs. 47 and 48, distinct dermal rays, *D*, are seen, extending from the body wall to the fin margin, splitting and segmenting distally in becoming more perfectly specialized in function. The cartilaginous supports, $R + N$ and $R + H$, must now be

---

* *Journal of Morphology*, IX, 1, 1894.
† Gephyrocercy of Ryder.

looked upon as including the **elements of both the radials and the** hæmal **or** neural processes and spines.

The Paired Fins

The paired fins of fishes claim an especial interest as the precursors of the limbs **of** the land-living vertebrates. **In this** light **they have been** widely studied, **and many** schemes have been **devised for** the comparison of the parts of the five-fingered extremity, or *cheiropterygium*, **of the** amphibian with the fin structures of many fishes. **The un-**satisfactory character of these homologies, **however, is felt at the present** time more **generally than ever, and** many morphologists **believe with Dr. Mollier** * that the ancestral form **of** the terrestrial limb cannot **be found in any of the** known types **of** paired fins.

Among fishes, on the other hand, **there appears to be a** well-marked unity **of plan** in **the varied forms of the** paired fins ; **and** there **exists so** perfect a gradation in structural characters in the different forms that it **seems** impossible to doubt **their genetic** kinship. **Which fin,** however, must be looked upon as the ancestral type **is still** disputed. Professor Gegenbaur **has long** maintained that the fin **of Fig.** 54 **(or,** better, the **pectoral** fin **of Fig. 147†)** is to be looked upon as the most primitive form, or *Archip-terygium.* It **is a** leaf-shaped **fin, whose** principal **carti-**laginous supports are arranged in a **row** from **base to tip in** the position **of a** mid-rib : and **whose minor fin supports are grouped more or less** symmetrically on either side of this axis (cf. Figs. **53, 54, 121, 123, 126).** The archipteryg-ium **is** believed **by Gegenbaur to** have had **a** centrifugal origin : **it arose** behind **the** gill region, representing **in its**

* *SB. Gesell. f. Morph. München,* 1894, p. **17.**

† Gegenbaur, *Das Flossenskelet der Crossopterygier.* Cf. *Morph. JB,* 1894.

supporting substance the fusion of the cartilages of the hindmost gill bars ; in its outward growth the median axis of the fin was first produced, the minor supports then arranging themselves on both anterior and posterior margins. The fin of Fig. 52 was believed to represent a specially evolved (or "monoserial") form of the archipterygium : the hindmost of its elements, *B*, was homologized with the primitive fin stem, along whose posterior (post-axial) margin the elements, *R*, no longer occurred.   The structures of Fig. 53 were adduced as a transitional stage in the differentiation of the biserial archipterygium (Fig. 54) into the monoserial form of Fig. 52.

The theory of Gegenbaur as to the origin and evolution of the paired fins cannot be said to be in any way generally supported at the present time.   The opposing view, that of their derivation from a continuous lateral dermal fin fold, based on the work of Thacher, Balfour, Mivart, Dohrn, Wiedersheim, and others, is widely accepted, and continues to gain supporting evidence on the sides both of embryology and palæontology.

In the following discussion of the paired fins the writer has mainly followed the recent studies of Wiedersheim.*

The paired fins are believed to have arisen as balancing organs, accessory in function to the vertical fins.   They probably occurred early in the line of descent as a response to a need for balancing the fish's body, at the time when the vertical fin was separated into caudal, dorsal, and anal elements.   There can be little doubt that they first arose in the line of the fish's motion, and are known primitively (Figs. 49, 50), as a pair of keel-like lateral lappets arising somewhat ventrally, and directed outward and downward.

* *Das Gliedmassenskelet der Wirbelthiere,* 1893.

The foremost **pair** appears **anteriorly not** far behind **the** gill region : from its position it has certainly the more **important** mechanical function in balancing the fish's **length** —on this account becoming more widely modified **in form** and function as the *pectoral* **fins.** The hinder pair, or **ventral** fins, though in the plane of **the** pectorals, has a more ventral position, **the** hinder borders converging in the region of the anus. **The** ventral fins **are** certainly placed in the most motionless region **of** the fish : they are little affected by either the lateral or upward movements of **the** body ; and remain accordingly smaller in size and simpler in structure than the pectoral **fins.** That there may have existed in primitive **fishes a third (post-ventral) pair of fins** is by no means **improbable (cf. T. J. Parker,** *Ref.* p. **244),** although its presence **has not as yet been satisfactorily** demonstrated.

The paired fins **thus appear to have been** derived from a continuous dermal **fold, similar in** every way to that giving **rise** to the vertical **fins.** They appear, moreover, to have undergone the same **mode of** evolution in their **structures as have the** dorsal or **anal fins.** The unpaired fin **fold as it** passed forward on **the ventral side of** the body **may** primitively have forked **in** the anal region, **and given rise on** either side to a lateral fold. **In** these might next appear an anterior and posterior pair of lappets, —pectoral **and** ventral fins,— whose positions would be determined **by** mechanical needs, and whose size would increase as **the** intervening and useless portion of **the** dermal fold **disap**peared. In the subsequent history of **pectoral and ventral** fins, supporting elements, actinotrichia, radials, **and basals,** would arise in the same way as in the unpaired fins, **and a** similar metamorphosis of the fin form would take place, owing **to the** concrescence of these elements **and** to the

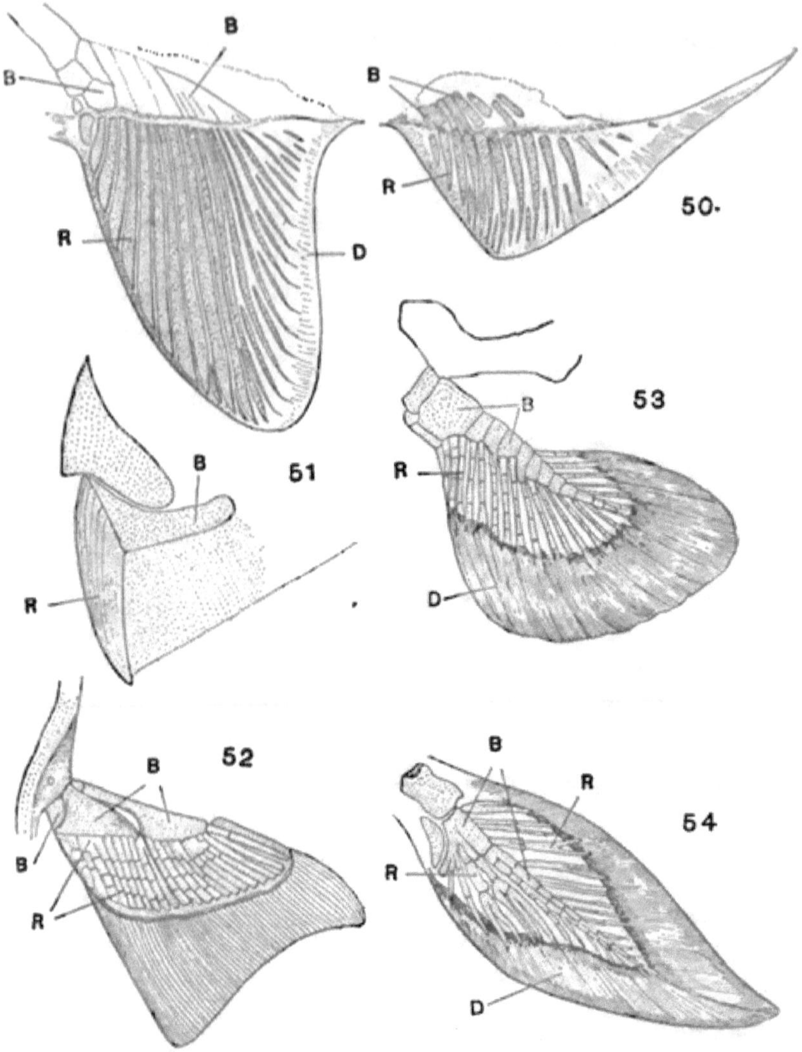

Figs. 49–54.— Evolution of paired fins. 49–50. Pectoral and ventral fins of Cladoselache. × 1. 51. Pectoral fin of Acanthodian, *Parexus.* (After Smith Woodward.) 52. Pectoral fin of *Heptanchus.* (After Gegenbaur.) 53. Pectoral fin of Xenacanthus (*Pleuracanthus.*) (After A. Fritsch.) 54. "Archipterygial" pectoral fin of Ceratodus. (After Howes.)

B. Basal. *D.* Dermal. *R.* Radial.

subsequent encroachment of the dermal fin margin. These
conditions may be briefly illustrated. The paired **fins of a**
primitive shark (Figs. 49, 50, v. **p. 79) appear as the actual**
lappet-shaped remnants of a continuous dermal fold. **The**
ventral fins (Fig. 50) have clearly retained even the outward shape of the fin fold; the supporting elements **are**
arranged in metameral order; the radials, R, are unjointed,
extending from body wall to fin margin; the basals, agree-
ing in number with the radials, are uniform in size, and **as**
yet unfused. The pectorals, acquiring more special functions (Fig. 49), are enlarged in size, their basals, B, **becom-**
ing compressed and obscure. **In these** fins the effect of
concrescence is admirably marked; the anterior fin **margins,**
pressed tail-ward in their plane of growth, become firm and
rigid, their elements stout and compact; the basals, re-
sponding to this outward need, cluster more firmly together,
are compressed and fused, their **anterior elements,** largest
and stoutest, become inturned, their **posterior** elements,
slightest and most clearly metameral.*

The **next** stage in **the** evolution of the paired fins **is**
clearly comparable to **that** already noted as occurring in
the dorsal **fin of** Holoptychius (Fig. 43), where the line of
basals, fusing compactly into a plate-like mass, had inturned its anterior, and protruded **its** posterior tip; **a**
change apparently slight, but great **in** functional importance. Up to this stage the fin has been firmly implanted
in the body wall; its motion, probably slight upward or
downward, served but to balance the fish, its fin rays,
tending to concentrate anteriorly, functioned **as an** efficient
cutwater. This process of concentration **in the** anterior
fin margin may have **resulted, the writer believes,** in the

---

* **The effect of** the enlarged and clustering dermal denticles in strengthening the cutwater margin of the fin has already been noted (p. 28).

formation of fin spine, as in Acanthodian * (Figs. 32, 51,
and p. 81).   But the protrusion of the line of the basals
must have brought with it a new use in the economy of
fish motion.   The plane of the fin could now be directed
upward or downward; the fin would become a direct aid
in propulsion; it would acquire a paddle-like function; it
could also be extended sideways as a check to motion.
Under these circumstances it is not unnatural that the
region of the concrescence of the fin rays should now be
transferred from the fin's anterior to the more useful posterior (now distal) margin, and that the fin rays, as well as
the line of basals, should acquire a more jointed structure,
suited to flexible motions.   The course of the differentiation of fin structures may be traced from this point onward, as Wiedersheim has shown, by means of a series of
gradational stages : from the conditions of Fig. 49 we may
in the present figures pass to those of Fig. 52, thence to
those of Figs. 53 and 54.   In the pectoral fin of a modern
shark (Fig. 52) the basal cartilages, *B*, may still be compared with those in the older form (Fig. 49 *B*); their distal
element (*B*, at the right of the figure), however, protrudes
from the body wall and is becoming surrounded by clustered radials, *R*; the cartilaginous elements, it is here
noted, have been placed in competition with the dermal
elements, and have already yielded them over half of the
fin area.   In the next stage of the evolution, as in the
pectoral fin of a Permian shark (Pleuracanthus, p. 83, Fig.
53), the line of the basals is seen to boldly protrude from
the body wall and to have become distinctly jointed; the
radials have surrounded its distal end, and taken a position

---

* This homology proposed by the writer has not been accepted by Smith
Woodward; the spine is unquestionably encased *outwardly* by dermal denticles.

along the outer half of the hinder margin of the fin stem ; the dermal region of the fin, *D*, has notably increased. Indeed, the fin area in the modern bony **fishes** (Fig 145, *PF*) may become entirely dermal, and the basal supports greatly reduced and metamorphosed. **In** a final type of fin (Fig. 54) the line of the **basals** has become widely specialized, and the characters of the archipterygium have been attained : the fin stem is long, tapering, jointed ; the radials occur as clearly along the hinder as along the **ante-**rior margin ; and, as in Figs. 52, 53, dermal rays **contrib-**ute largely to the fin area. This form of fin may be noted **as most** closely approximating in function the **limb type of** land-living vertebrates.

It has recently **been urged that** the lateral fold origin **of** the paired fins as thus described is **not** confirmed by developmental studies, — the especial **ground for this** belief being that in sharks these fins appear, **even in** very early stages, as paired lappet-like outgrowths, destitute of intervening fin membrane. The perfected fin fold is therefore claimed to represent nothing more than a specialization to bottom-living, since this condition is known to maintain in earlier stages and in more primitive metamerism **in the** development of skates : and as skates (p. 93) are well known to represent a comparatively recent offshoot from the stem of the sharks, it is accordingly inferred that the chief **proof** of the lateral fold doctrine is destroyed.

Since these objections, however, were raised, the structural conditions **of** the ancient shark **of** Figs. **49 and 50** have been described, **and** may be looked upon as **the** weightiest evidence **of the** origin **of** paired **fins from** lateral folds. Nor **does it** seem to the present writer that the early character **of the** fin-fold metamerism of skates is to be looked upon **as** an unexpected condition. Their broad longitudinal fins, specialized to bottom-living, become fashioned in an ancestral **mould** ; and it seems not unnatural that **they tend to** reacquire their latent primitive form **at an early** period.   **On the** other hand, the fin-fold condition **of** the shark might **be** less perfectly shown on account of processes of **accelerated development.**

4. THE CHARACTERS OF THE SENSE ORGANS OF FISHES

It has already been seen that the conditions **of** aquatic **living have caused** fishes to evolve adaptive structural **characters, such as body form, specialized** metamerism, organs **of progression, and dermal investiture.**   It is not, accordingly, **unnatural to expect that, from the same** causes, the **condition of the sense organs may have been** strikingly modified.

The sense of "feeling" — using **the word in** its general meaning — **has been** of **especial value in fishes,** and tactile **organs appear to be** independently **developed** in all fish **groups whose living habits demand them.   In the** form of *barbels* **they thus occur in members of the various** divisions **of bony fishes, as cod (cusk,** *Ophidium*) (Fig. 55), **drum-fish,** *Pogonias* **(Fig. 56), or** sculpin, *Hemitripterus* **(Fig. 57).   Their form may be lobate, thread-like,** or villose ; **they are often surprisingly similar in** size, position, and **innervation ; they usually appear on the** inferior head **surface, most often in the anterior** throat region, in the **position most exposed to tactile** impressions.   The threadlike **barbels of the catfishes (Fig. 58, p. 171)** are arranged **in pairs about the margin of the** mouth ; **the** longest lateral **pair is connected with the marginal bone** (maxillary) **of the upper jaw and** directed at will.   **In other** mud-living **forms, sturgeons** (Fig. 160), the barbels have arisen on the **under side of the shovel-like snout,** directly in advance of

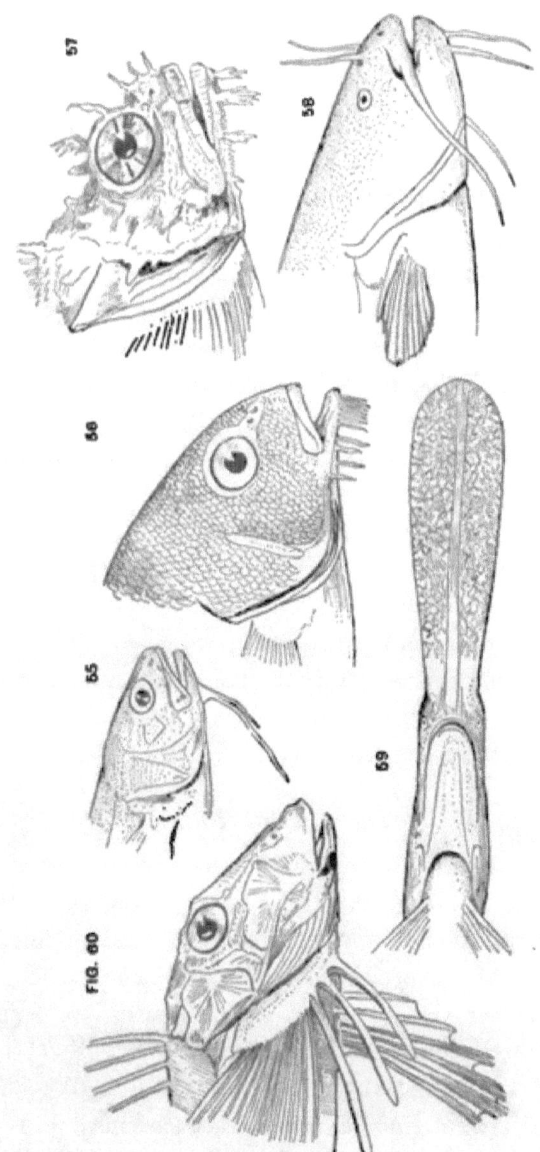

Figs. 55–60. — Barbels and tactile sense organs. (After GOODE in U. S. F. C.)
55. Cusk, *Ophidium*. 56. Drum-fish, *Pogonias*. 57. Sea-raven, *Hemitripterus*.
58. Catfish, *Amiurus*. 59. Spoon-bill sturgeon, *Polyodon* (ventral view of snout).
60. Sea-robin (Gurnard), *Prionotus*.

the protractile sucking mouth. There can **be** little doubt
that the most aberrant tactile **organ in** fishes is the long
spatulate rostrum of the paddle-fish (Polyodon) of the Mis-
sissippi (Fig. 59) : the sense organs are here known **to be
most highly specialized,** although their intimate structure
is as yet not understood. **Tactile** organs are often to be
found upon fin structures, especially those of the anterior
body region. In the sea-robin, *Prionotus* (Fig. 60), the sen-
sory structures are borne by three anterior fin rays ; these
are greatly enlarged, lose their connecting fin web, and
can be moved at will in a variety of ways. In all cases
the barbels appear. to be true and highly specialized
organs **of touch, and the end** organs are comparable apparently with the touch papillæ of higher forms. Of their
extreme sensitivity there can be no doubt, and as far as
can be judged from their innervation, it would appear that
their function is tactile rather than gustatory, as has been
suggested. The limits of these processes, however, are
no doubt poorly defined in aquatic living.

The Lateral Line

The sense organs, generally known as the *lateral line,*
or *mucous canal system,* **are looked upon as** essentially
peculiar to fishes. In the form of a 'lateral line,' they
are arranged more or less segmentally along the median
line of either side of the body and form a conspicuous
feature in the outward appearance of the fish (Figs. 87,
104, *LL,* **121,** *LL,* **145,** *LL*). Often by striking colora-
tion, the lateral line is rendered even more prominent,
passing from the head to the tail as a pale or brightly
coloured band, against the dusky side of the fish. In the
region of the head, however, this sensory structure is, as
a rule, no longer conspicuous : it dips below the skin sur-

face and becomes a series of interconnecting tubes, **which pass** along the most exposed ridges of forehead, **cheek,** orbit, and jaw **rim.** Here in different regions, **these sensory** mucous tubes may become dilated, constricted, **or** ramose, and may communicate with the surface by occasional or numerous pores.

The mucous canal system **has long** been a subject **of** study and investigation. It is looked upon generally **as a** sensory organ, adapted to the conditions of aquatic **living,** but its function has not been definitely established. **How** it was acquired, or how its ancestral conditions **have been** modified in the present groups **of** fishes, must at **present** be looked upon **as in many ways** doubtful.

The **simplest conditions of the mucous canal system** appear to **exist in** primitive **sharks: and to these the** writer believes that the **modified sense canals in other** fishes may best be referred.

The ancestral condition of **the** lateral line **of** sharks appears to have been represented in an open continuous **groove,*** lined with ciliated sense cells, and protected **only by an** overcropping margin of shagreen denticles (Fig. **61**). **In** this condition **it at** least **exists in the** ancient sharks of Figs. 86, 87, **92, and** in the Chimæra (Fig. 104). That the canals of **the head** region were also primitively of this character appears exceedingly **probable:** they are thus retained in the adult Chimæra (Fig. 104, *M.C*).†

In the modern forms of sharks **the condition of the**

---

* It is to be noted that this condition occurs in deep-sea fishes: it **here is** evidently an adaptation to their peculiar environment, which causes an **early** ontogenetic stage to be permanently retained.

† In Callorhynchus this condition **has been largely lost :** the outer margins of the sensory groove **have sealed over.**

E

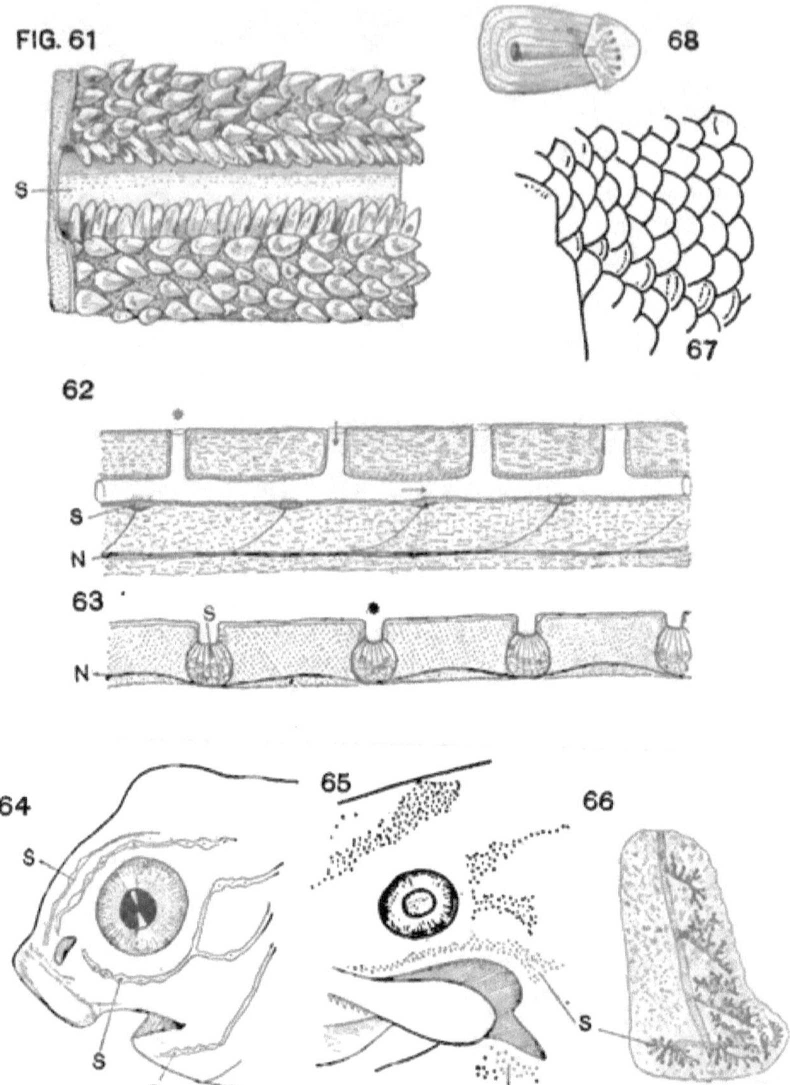

Figs. 61–68. — Mucous canals (lateral-line organs). 61. *Chlamydoselache*, groove-like lateral line. (After GARMAN.) 62. Plan of lateral line of sharks, longitudinal section. 63. Plan of sensory end buds (lateral line). 64. Sensory tracts of head of larval Amia. 65. Surface openings of tubules of sensory tracts of head of adult Amia. 66. Ramification of sensory tubules in dermal plate of Amia. 67. Cycloid scales of Amia, showing the openings of the tubules of the lateral line. 68. Cycloid scale of the lateral line of Amia, showing the course of the sensory tubule. (Figs. 64–68 after ALLIS.)

N. Nerve supply. *S.* Sensory tissue. * Denotes an outer opening; → the direction of an incoming stimulus.

sensory canals suggests the modifications to which the open sensory groove has been subjected. There are thus forms in which the canal becomes more and more deeply sunken in the integument, and acquires a tubular character by the fusing together of its outer margins. The section of the lateral line of the Greenland shark, *Læmargus* (Fig. 62, v. p. 90), shows the tube-like sensory canal well sunken from the surface, but retaining metameral openings at the points. The sensory cells, S, are no longer, as in Fig. 61, scattered evenly along the floor of the canal; they now occur in metameral masses supplied with a distinct nerve branch, N, located in the region immediately below the external tubules. When sunken in the integument, the sensory canal is known to have acquired supporting structures to enable its tubular character to be maintained; in the Cretaceous shark, *Mesiteia*, an elaborate series of surrounding calcified rings [*] were thus evolved.

Further changes in the mucous canal are often accompanied by the subdivision of the external apertures; each of the openings of Fig. 62 might by this process give rise to a series of minute surface pores, as at S in Fig. 65, or enlarged, showing the collecting mucous canals in Fig. 66. This ramose mode of termination of the external tubules has been admirably described by Allis [†] in the ontogeny of a ganoid; in a larval stage (Fig. 64, S, S, S), the condition of the sensory canals is seen to differ little from those shown in section in Fig. 62; although imbedded in the integument, occasional pores are seen, S, S, to open to the surface; these subsequently by repeated subdivision give rise to the great number of minute open-

[*] A condition somewhat similar has been noted (Leydig) in Chimæra.
[†] On the Lateral Line System of *Amia calva*. *J. of Morph.*, 1889.

ings already noted in Fig. 65. A process of this kind is carried to great lengths among the fishes which develop horn-like scales, as Amia, herring, or cod: in the scales of the lateral line region the distal tubules appear at the surface as a cluster of pores, as shown in Fig. 67, or in the detached scale of Fig. 66.

The organs of the lateral line (of a bony fish) shown in section in Fig. 63 are regarded by the writer as of a highly modified character. They appear to have been derived from the conditions of Fig. 62; the end organ, *S*, corresponds with that, *S*, of the preceding figure; its size, however, has greatly increased, and the intervening sensory tube has been lost; its metameral opening at the surface corresponds with that of Fig. 62; the nerve supply, *N*, is now seen to have secured a more perfect relation to the end organs.

The original significance of the lateral line system as yet remains undetermined. As far as can be judged from its development, it appears intimately, if not genetically related to the sense organs of the head and gill region of the ancestral fish: in response to special aquatic needs, it may thence have extended further and further backward along the median line of the trunk, and in its later differentiation acquired its metameral characters.

A significant feature of its development is its peculiar innervation. Its lateral tract is innervated by a specially evolved root of the vago-glossopharyngeal group, but its head region is supplied by a similar root of the facial nerve (perhaps also by the trigeminus; cf. Collinge, *Ref.* p. 248).

In view of this innervation, the precise function of this entire system of end organs becomes especially difficult to determine. Feeling, in its broadest sense, has safely been admitted as its possible use. Its close genetic relationship with the hearing organ suggests the kindred function of determining waves of vibration. These are transmitted in so favourable a way in the aquatic living medium, that from the side of theory a system of hyper-sensitive end organs may well have been specialized. The sensory tracts along the sides of the body are certainly well situated to determine the direction of the approach of friend, enemy or prey.

The Pineal Eye

The presence or absence in fishes of the *pineal end organ*, the "unpaired median eye of chordates," may finally be noted, since the condition of the *epiphysis* and its associated structures in fishes has an important bearing on general vertebrate morphology.

It is well known that in many forms of reptiles there exists, at the distal end of the epiphysis, a well-defined sensory capsule, whose structure shows unquestionably its optic function. It has seemed to many, therefore, that throughout the chordates the epiphysis has been primitively associated with a median eye, which has degenerated as the paired eyes became better evolved. That it has been retained in an almost perfect condition in reptiles has accordingly been looked upon as an outcome of a life habit which concealed the animal in sand or mud, and allowed the forehead surface alone to protrude:—the median eye thus preserving its ancestral value in enabling the animal to look directly upward and backward.

If this view as to the presence of a parietal eye in the ancestral vertebrate is to be generally accepted, one would naturally suggest that the organ should be present, at all events to a recognizable degree, in some of the varied forms of the lowest vertebrates extant, —fishes and amphibia. If there are no suggestions of its visual nature among these forms, one would be inclined to believe with O. Hertwig, that the epiphysis was originally of a different function and that its connection with a median eye may have been altogether of a secondary character.

The evidence as to the presence, primitively, of a median eye in fishes is certainly far from satisfactory : * in all the forms of recent fishes, no structure has been found associated with the epiphysis which, by the broadest interpretation, could be looked upon as suggesting a visual function. It is possible that fishes and amphibia may, in their extant forms, have lost all definite traces of this ancestral organ on account of some peculiar condition of their aquatic living. On this supposition, evidence of its presence might be sought in the pineal structures of the earliest Palæozoic fishes — whose terrestrial kindred, and probable descendants, may alone have retained the living conditions which fostered its functional survival.

It is accordingly of interest to find that in a number of fossil fishes the pineal region retains an outward median opening, whose shape and position suggest that it may have enclosed an optic capsule. If the median eye existed in these forms, it may well have been passed along in the line of descent through the early amphibia (where substantial traces of a parietal foramen occur, *e.g.* as in Cricotus) to the ancestral reptiles. This view is greatly strengthened, as Beard has shown, by the presence in the lamprey of a pineal end organ (optic?).

The evidence, however, that the median opening in the head shields of ancient fishes actually enclosed a pineal

---

* Hertwig (Mark), *Handbook of Embryology of Vertebrates*, and Cattie, *v. Ref.* p. 250.

eye, is now felt by the present writer to be more than questionable. The remarkable pineal funnel of **the Devonian** *Dinichthys* (Fig. 134) is evidently to be compared with the median foramen of *Ctenodus* and *Palædaphus* (= Sirenoids, p. 122) ; but this **can** no longer **be** looked upon as having possessed an optic function, and thus **practically** renders worthless all the evidence of a median eye presented by fossil fishes. It certainly appeared that **in the** characters **of the** pineal foramen of Dinichthys there **existed** strong grounds for believing that a median **visual** organ was present : its opening was in the **pineal** plate, midway between the orbits (*PN*, Fig. 134). At the surface **it was** of minute size (*X*, Fig. 136), but **below (Fig. 137) it flared** out into **a** funnel-like form, shown **in longitudinal section in** Fig. **137 *A*.** The peculiar character of this opening seemed to render it especially fitted for **a** visual function ; the minute external opening **forms** an image near the plane of the visceral opening of the funnel, without the specialization of a lens, — an **image so perfect that** it might readily be photographed. It is evident, accordingly, that if an optic capsule were enclosed **by this foramen,** it would have enabled its possessor to **have looked** directly upward and backward; and, **without the need of** developing lens-like and focussing structures, it could have readily received the images of all outer objects **near or** remote.

But the function of this pineal foramen, unfortunately for speculation, could not have been optical. **It occurs in a fish** (*Titanichthys*) closely related to Dinichthys, and, as the writer * **has recently found, is** *of a distinctly* **paired**

---

* He is obliged by accumulating **evidence to abandon his former view that** the pineal foramen of Dinichthys contained a **specialized optic capsule** (*N. Y. Rep. of Fisheries*, 1891, pp. 310–314).

character, its visceral and outer openings bearing grooves and ridges which demonstrate that the pineal structures must not only have been paired, but must have entered the opening in a way which precludes the admission of the epiphysis. It is now, therefore, that the pineal foramen which has been described in Siluroids * becomes of especial interest, since its contained structures are apparently connected with the lateral line system of paired nerves.

It must for the present be concluded, accordingly, that the pineal structures of the true fishes do not tend to confirm the theory that the epiphysis of the ancestral vertebrates was connected with a median unpaired eye ; it would appear, on the other hand, that both in their recent and fossil forms, the epiphysis was connected in its median opening with the innervation of the sensory canals of the head. This view, it is now interesting to note, seems essentially confirmed by ontogeny. The fact that three successive pairs of epiphysial outgrowths have been noted in the roof of the thalamencephalon, appears distinctly adverse to the theory of a median eye.

* Dean, *N. Y. Rep. of Fisheries*, 1891, and Klinckowström, *Anat. Anz.*, 1893, viii, p. 561.

III

THE LAMPREYS AND THEIR ALLIES

THE relations of the more primitive chordates to the true fishes have not been considered in the present discussion. A brief account, however, must be given of the Cyclostomes, or Marsipobranchii, which are represented in the recent lampreys and hags.

The three prominent forms of Cyclostomes are figured on a following page (Figs. 70–72, *A–D*). They are eel-like in shape, but are lacking both in paired fins and in an under jaw. Their mouth is of a rounded form, and is suctorial; when closing, its lateral margins draw together. Their skeleton is of the simplest character, membranous rather than cartilaginous; its elements are never more highly differentiated than those shown in the accompanying figure (Fig. 69, *A*).

Bdellostoma is shown in surface view in Figs. 70 and 72 *A*, and in sagittal section in Fig. 69. It is looked upon as the most archaic form of the living Cyclostomes. Barbel-like structures surround its mouth region; its nasal canal (Fig. 69, *N* and *C*) has a forward opening at the snout, and a hinder one piercing the roof of the pharynx, —a very exceptional character in fishes; its tongue, studded with rows of rasp-like teeth,* may be greatly everted,

---

* The teeth of Myxinoids are cuticular structures, and may well have been evolved within the limits of the group. Beard has homologized them with the teeth of sharks, but his determination of the presence of true enamel has not been confirmed (Ayers).

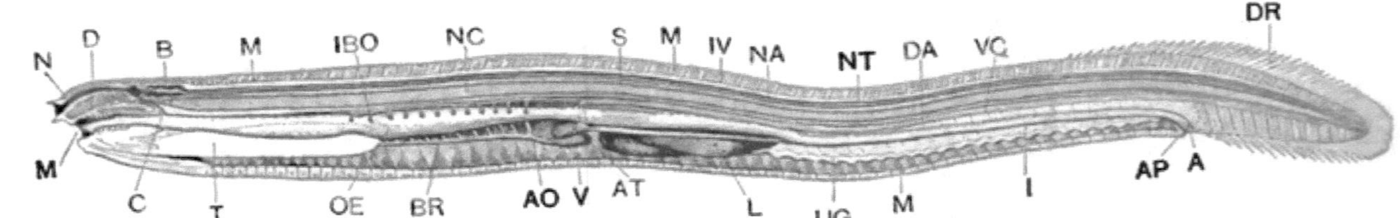

Fig. 69. — General anatomy of Cyclostome, *Bdellostoma dombeyi*, shown in sagittal section.

A. Anus. *AO.* Afferent aorta. *AP.* Abdominal pore. *AT.* Atrium. *B.* Brain. *BR.* Gill pouch. *C.* Canal opening from nasal organ to roof of pharynx. *D.* Dental cusps. *DA.* Dorsal aorta. *DR.* Dermal fin rays. *I.* Intestine. *IBO.* Inner branchial opening. *IV.* Septum between muscle-plates. *L.* Liver. *M.* Muscle-plate. *N.* Nasal canal. *NA.* Sheath of spinal cord. *NC.* Notochord. *NT.* Spinal cord. *OE.* Gullet. *S.* Sheath of notochord. *T.* Tongue (in dotted lines). *V.* Ventricle. *UG.* Genital fold (containing reproductive organs near ventral rim, and near its base the long kidney (pro- and mesonephros) and its ducts). *VC.* Right cardinal vein.

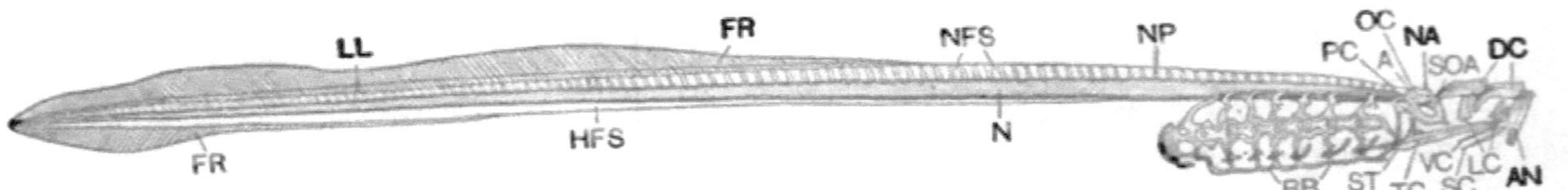

Fig. 69 A.—Skeleton of Cyclostome, *Petromyzon.* × 1. (In part after T. J. PARKER.)

A. Auditory capsule. *AN.* Annular cartilage surrounding mouth. *BB.* Cartilages of branchial basket. *DC.* Dorsal cartilages of mouth region. *FR.* Dermal fin rays. *HFS.* Fibrous sheath of blood-vessel. *LC.* Lateral cartilages of mouth region. *LL.* Longitudinal ligament, connecting apices of neural processes. *N.* Notochord (including its sheath). *NA.* Opening of nasal capsule. *NFS.* Fibrous sheath of spinal cord. *NP.* Neural processes. *O.* Openings of gill chambers. *OC.* Occipital crest. *PC.* Parachordal cartilage of cranium. *SC.* Styliform cartilage. *SOA.* Suborbital arch. *ST.* Styliform cartilage of suborbital arch. *TC.* Tongue cartilage. *VC.* Ventral cartilage.

as in Fig. **72**, *A*, and then drawn in by stout **tongue** muscles, *T* (**Fig.** 69) ; its digestive tube is almost straight, terminating at the base of the tail region at *A* ; **the** region of the gullet, *OE*, is pierced **by a** number **of** branchial openings, **varying** from seven **to** fifteen, **often** assymmetrical. The body cavity **is an** extremely **large** one for the size of the contained **viscera.** An unpaired fin, supported by delicate, unbranched **(dermal) rays is** restricted to the hindmost part of the body. Passing down the side is a row of mucous pouches by which a remarkable supply of slime is secreted. The living animal is enabled, by the peculiar character **of this** slimy **secre-** tion, to render a pailful of water jelly-like **in consistency.**

Bdellostoma occurs plentifully in **the bays of the Pacific** coast of America, **notably** at Monterey, **California. It is active** in its movements, **is carnivorous, and is well known** to take a baited hook. Its numbers make it an enemy of the fishermen, entangling and sliming their set lines, and destroying the captured fish. **It** is said to feed at night, although little is yet **known of** its general habits of living. None but adult specimens have thus far been observed.

The **Hagfish,** *Myxine glutinosa* (Fig. 71, and 72, *B*), is in many regards similar **to** Bdellostoma ; **it** differs mainly in the character of its unpaired fin and in **its** branchial **struct-** ures (Figs. 9, **10**). As already noted, the outer ducts **of the** gills, instead **of** opening separately at the surface as in Fig. 70, **are** drawn together tail-ward, and terminate **on** either side in a common ventral opening (**Fig.** 71, at the point *). The unpaired fin is almost lacking in supports ; its ventral origin is **even as far forward as the** branchial openings ; the anus, **as a** slit-like opening, pierces it in the tail region. As in Bdellostoma, the nasal canal begins at the snout, and at its hinder opening pierces the roof of

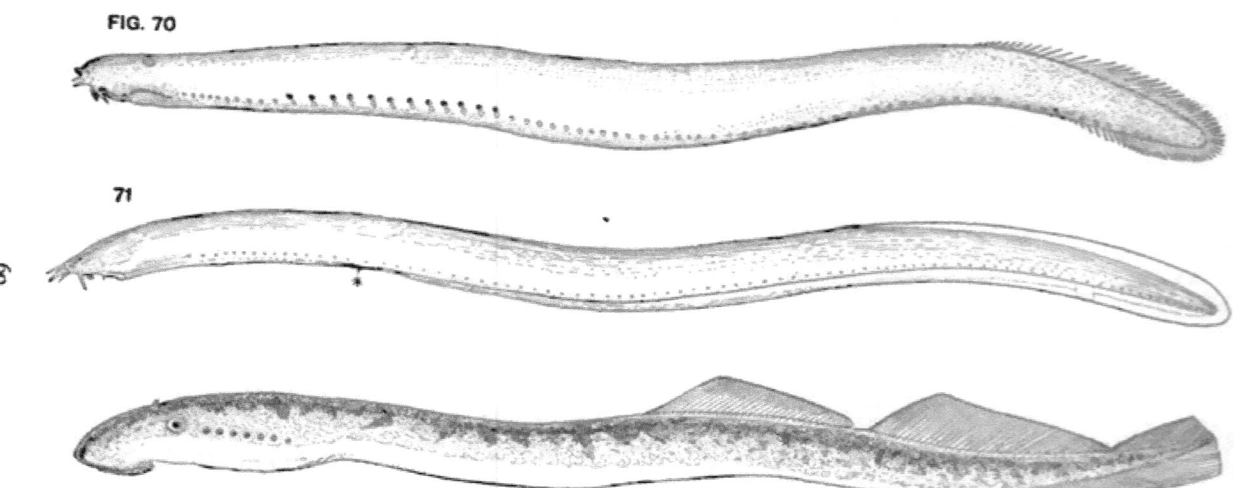

Figs. 70-72. — Cyclostomes. 70. *Bdellostoma dombeyi*, Lac. Chile. The light apertures along the side are mucous pits, the dark ones branchial openings. 71. *Myxine glutinosa*, L. The light dots along the side are mucous pits; the left common branchial aperture is at *. 72. *Petromyzon marinus*, L.

the pharynx; this, with other related conditions, has caused Myxine and Bdellostoma to be included in a sub-group of Cyclostomes, as Myxinoids, or *Hyperotretes*.* In each genus there is possibly no more than a single valid species.

Myxine is a well-known form: it occurs along the Atlantic coast at moderate depths. It is exclusively carnivorous,

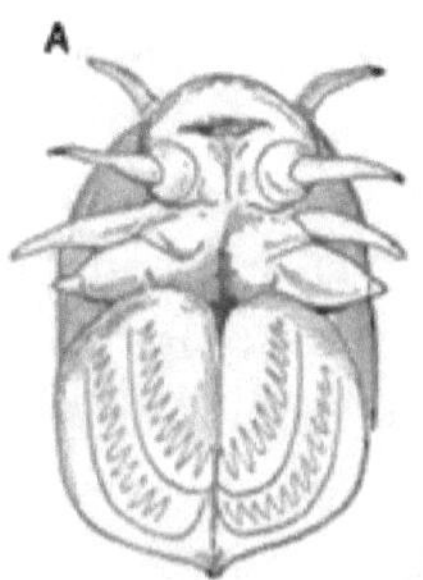
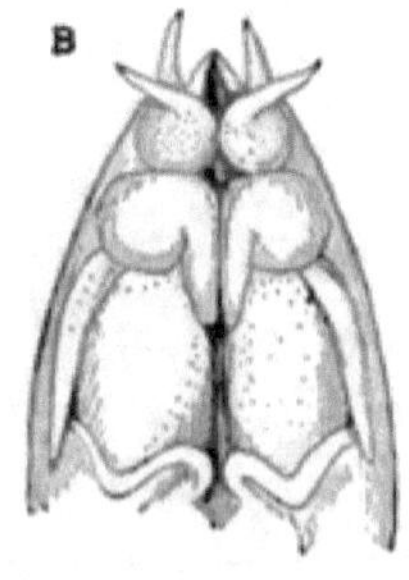
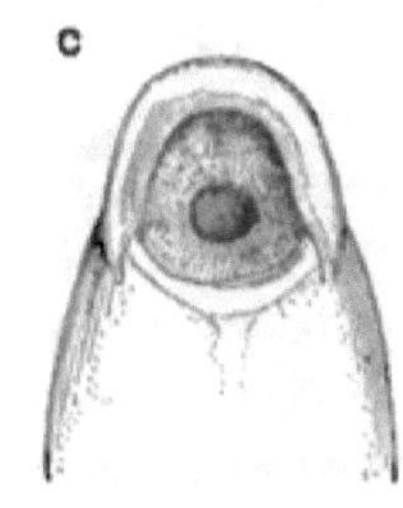
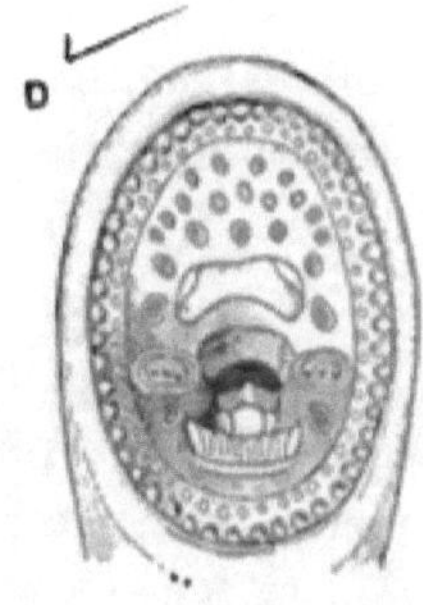

Fig. 72.—*A–D.* Ventral aspects of heads of (*A*) *Bdellostoma* (after AYERS); (*B*) *Myxine* (after GÜNTHER); (*C*) *Ammocœtes* (after GÜNTHER); (*D*) *Petromyzon* (after GÜNTHER).

often boring its way into the abdominal cavity of (diseased or injured) fishes, and with them is brought to market; it is also taken not infrequently by line fisher-men. The smallest example that has thus far been described is 6 cm. in length; it was recorded by Beard. (V. *Ref.* p. 239).

The Lamprey, *Petromyzon*, is the most perfectly studied member of the Cyclostomes. Its species are common to the continents of the northern hemisphere; and in South America and Australia there occur very closely allied genera, as *Mordacia* and *Geotria*. The largest lamprey, *P. marinus* (Fig. 72, and *C, D*), is known to attain a length of nearly four feet; it occurs in the coast

* v. Glossary, p. 228.

rivers, ascending them in numbers in the springtime
(April) on the way to the spawning grounds (v. p. 182).
During its adult life it is supposed to be exclusively carnivorous, to some degree, perhaps, parasitic, although many
doubt that it is truly parasitic in the sense of entering the
body cavities of healthy fishes. It certainly is often taken
attached to other fishes, as shark, sturgeon, or salmon.

Immature lampreys differ so strikingly from the adults
that they were formerly regarded as species of a separate
genus, *Ammocœtes* (v. p. 215). In feeding habits the ammocœte is widely unlike the mature form ; it is toothless
(Fig. 72, *C*), and in part mud-eating, *i.e.* vegetivorous.

Petromyzon must be regarded as the most highly organ-
ized of Cyclostomes. Its mouth has no longer the fringing barbels of Myxinoids, — which suggest, according to
Pollard, the buccal cirrhi of Amphioxus, — it has acquired
stout supporting cartilages and a funnel-shaped form,
studded with a series of conical teeth, as shown in Fig.
72, *C*. The teeth of the hinder mouth region now appear
almost as though they were supported by a mandibular
cartilage ; the tongue, as in other Cyclostomes, bears the
teeth which are probably of the greatest functional importance. The nasal canal of Petromyzon has its outer opening
on the dorsal surface of the head ; its inner end, however,
does not perforate the roof of the mouth, although produced
backward as a blind sac, closely apposed to the pharynx.
Petromyzonts are, accordingly, arranged as the sub-group
Hyperoartia, in contrast to the Myxinoids.

Further structural characters, which the lamprey seems
to have derived from simpler conditions, may be noted in
its unpaired fin, gill chamber, nervous system, and skeleton. The unpaired fin has subdivided into dorsal and
caudal elements, and is now supported by well-marked

rays, which (sometimes) bifurcate. The branchial region of the adult lamprey's gullet **is** restricted to **a** pouch-like diverticulum (v. p. 263 and Fig. **326**). A 'sympathetic' nervous system, and a **'lateral line' has appeared : the** latter passes down the side **in two** branches, one **above** and one below the median **lateral** plane : its end organs are the pouches of nervous epithelium which in **Myxi-** noids are scattered generally over the body surface. **The** skeletal structures of the lamprey (Fig. 69, *A*) indicate well-marked advances : a stouter supporting tissue of car- tilage-like character has appeared ; the brain **case** is partly roofed over ; neural processes, *NP*, a branchial basket, *BB*, **and a series of mouth** cartilages **are especially note-** worthy.

Affinities of the Cyclostomes

The relations **of the group, Cyclostomi,** to the earlier chordates, and, on **the other hand,** to fishes, have been by **no** means definitely established. Dohrn and others have suggested that the Cyclostomes are greatly degenerate, and are even closely akin to the recent bony fishes, as perch or cod. Their views have been based upon several struct- ural characters, notably vestigial organs, such **as** the ap- pendages at the sides of the cloacal opening of Petromyzon which were believed to represent pelvic fins ; and there was further taken into consideration the belief that the entire group was one of degenerate life habits. The views of these writers, however, do not appear to be confirmed by later studies, and the belief is becoming more and more general that Cyclostomes represent a **very** ancient chordate stem whose ancestral form is most nearly exemplified **by** Bdel- lostoma. Parasitism has been acquired to a limited degree, but does not appear to have affected the general characters of the group. Among its primitive features are to be included : **skeleton and** muscles, continuous vertical fin, gill **characters (p. 260), viscera (p.** 263), urino-genital organs **(pp. 266, 270), nervous and** circulatory systems (pp. 260, **269, and 274). With these must be** taken into account: absence of mandible * and of paired fins and girdles; and in **addition** the remarkable conditions **of metamerism (p.** 14).

Little more that a vague kinship **between** lampreys and fishes **has been established by the study of** living forms. **And, on the other hand,** it would appear equally impracti**cable to** obtain **evidence bearing upon** this problem from **the** side of palæontology. **All that is known** of the recent Cyclostomes more than suggests that their soft body struct**ures would prove most** unfavourable **to** fossilization. It **would be only, therefore, in the event of** some of their **ancient members** possessing calcified structures that palæ**ontology would be** able to offer **a clue** as to their ancient affinities.

Upon the problem of their descent the evolution of **fishes has, however, an undoubted** bearing, in suggesting **the lines and** effects **of aquatic** evolution and the perma**nence** of generalized **types. It certainly** tells of the ex**treme slowness of the evolution of** aquatic forms and con**vinces us that the ancestral Cyclostome** could only have **occurred in a time stratum** exceedingly remote. Palæon**tology cannot perhaps hope to** obtain more than sugges**tions of the ancestral forms, although** these, from their **generalized characters, may well** have survived during geo

---

* **The cartilages of** the mouth region of Cyclostomes **have been** homologized **with the** structures **of** gnathostomes ; Pollard recently (*Anat. Anz.* ix, pp. **349–359)** ascribes a cirrhostomial origin to the mouth parts of a Teleostome **(catfish), which the writer** cannot believe has been demonstrated; variations **in the number, shape, and function of the cartilages of** the mouth rim of Cyclostomes might **well have occurred** within the limits of this ancient group.

logical ages. It can, however, show that Cyclostomes are not the degenerate descendants of shark-like forms ; and — if only by analogies in the evolution of fishes — it may still be able to demonstrate with fair probability their genetic kinships. It may, for example, prove that in the most ancient time there existed undoubted Cyclostomes, and that these in many and most specialized forms were even then branching-off twigs of a great descent tree. In such an event an inference would certainly be the more reasonable which derived the advancing line of fish descent from the genealogical tree of the more primitive Cyclostomes, than that *vice versa*.

It is now accordingly of especial interest that the fossil remains of what seems undoubtedly a lamprey (Fig. 73) have been discovered in the Devonian ; and this, together with a better knowledge of the ancient and curious chordate group, Ostracoderms, may, it is hoped, lead to some solution of the Cyclostome puzzle.

Fig. 73. — The Devonian Cyclostome, *Palæospondylus gunni*, T. × 4. (After TRAQUAIR.) Achanarras quarry, N. Scotland.

The Ostracoderms

Ostracoderms, as they are called from their shell-like, dorsal and ventral derm plates, are certainly the oldest known remains of vertebrates.[*] In their simpler forms they occur in the Upper Silurian ; they flower out in a variety of types in the Devonian, and shortly become extinct. In the present con-

---

[*] The earlier (Ordovician) vertebrate remains described by Walcott are as yet uninterpretable.

F

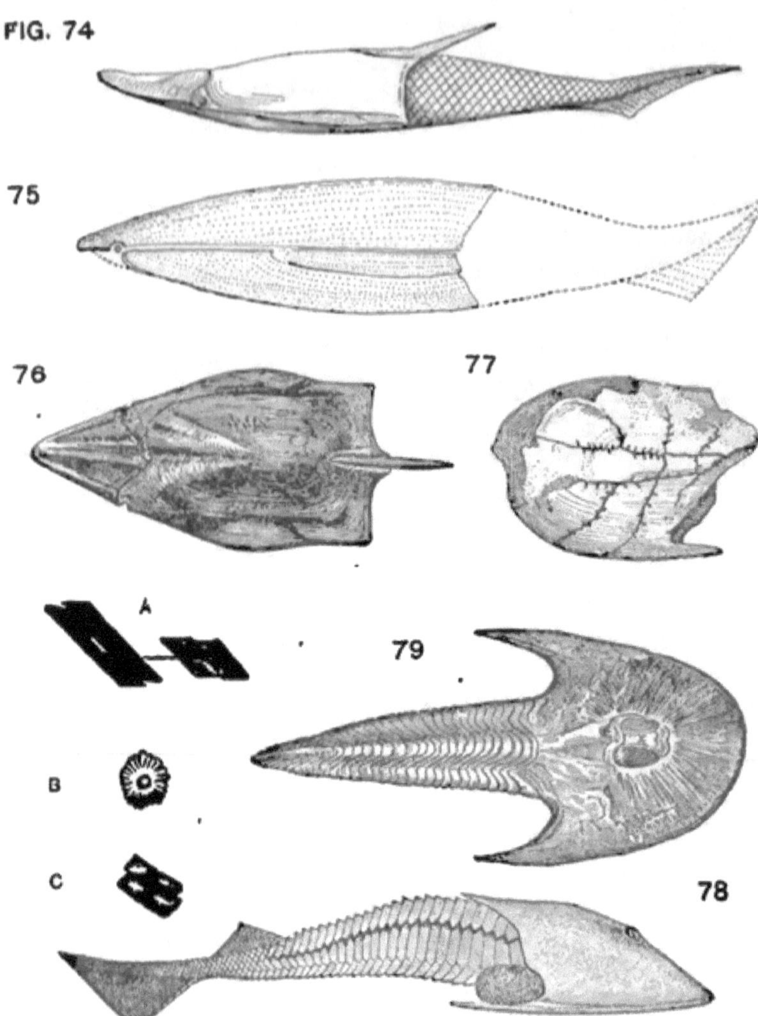

Figs. 74-79. — *Pteraspis* (restored). × ⅓. (After LANKESTER.) Lower Old Red Sandstone, Herefordshire. 75. *Palæaspis americana*, Claypole. × ⅓. (Restoration after CLAYPOLE, somewhat modified by the writer.) 76. *Pteraspis*, dorsal shield, slightly restored. (After LANKESTER.) 77. *Pteraspis*, ventral shield ("*Scaphaspis*"), showing mucous canals. (After SMITH WOODWARD.) 78. *Cephalaspis lyelli*, side view. (Restored by LANKESTER.) 79. *Cephalaspis lyelli*, dorsal aspect. × ⅓. (After L. AGASSIZ.) Specimen from Old Red Sandstone, Forfarshire. *A C.* Rhomboidal scales from different body regions. *B.* Tessera from middle layer of head shield.

nection they **may be** described, if only to **indicate** that they are in no way closely connected with **the** ancient shark types (p. 78), and that they are accordingly of but indirect interest in the descent **of** jaw-bearing vertebrates.

Ostracoderms may readily **be** reduced to **three general** types, *Pteraspid*, **Cephalaspid**, and *Pterichthid*. The first, oldest, and probably simplest occurs **in the Lower Old Red Sandstone of** Herefordshire. It was provided **with** arched back **and** breastplate (Figs. **74, 76, 77**), from **whose** anterior lateral notches a pair **of** eyes protruded ; **the surface** of these plates (Fig. **77**) appears to have been grooved for sensory canals. Pteraspis, **as seen in the** restoration, had a snout plate, **a** dorsal spine, **and a** body casing of rhomboidal scales ; **its** mouth was probably **in the** region immediately below **the eyes, in** front of the margin of **the** well-rounded ventral **plate ; this was generally regarded** as the dorsal plate of a kindred genus, "*Scaphaspis.*" Closely related is the American **Pteraspid**, *Palæaspis* (Claypole), from the Upper Silurian **of Pennsylvania (Fig. 75)** ; this form lacks the dorsal spine **of the** English species ; **it has a** well-marked lateral plate intervening between those **of the** back and ventral side, and, according **to its** discoverer, Professor Claypole, **possessed pectoral** fins **similar** to those seen in Fig. 123. **Its hinder trunk region is unknown.**

Cephalaspis, the second type **of** Ostracoderm, **is** from the Old Red Sandstone of Scotland (Figs. 78, **79**). **It was** curiously suggestive of a trilobite, and with little **doubt** mimicked this ancient crustacean **in its** life habits. **Its most** prominent feature is a crescent-shaped head, **with** sharp rounded margin like **a** saddler's knife. This is protected dorsally **by** but **a** single plate, arching upward and backward; at **its** summit was a pair of closely apposed eyes, **and** near its flattened **rim** were pouch-like sensory

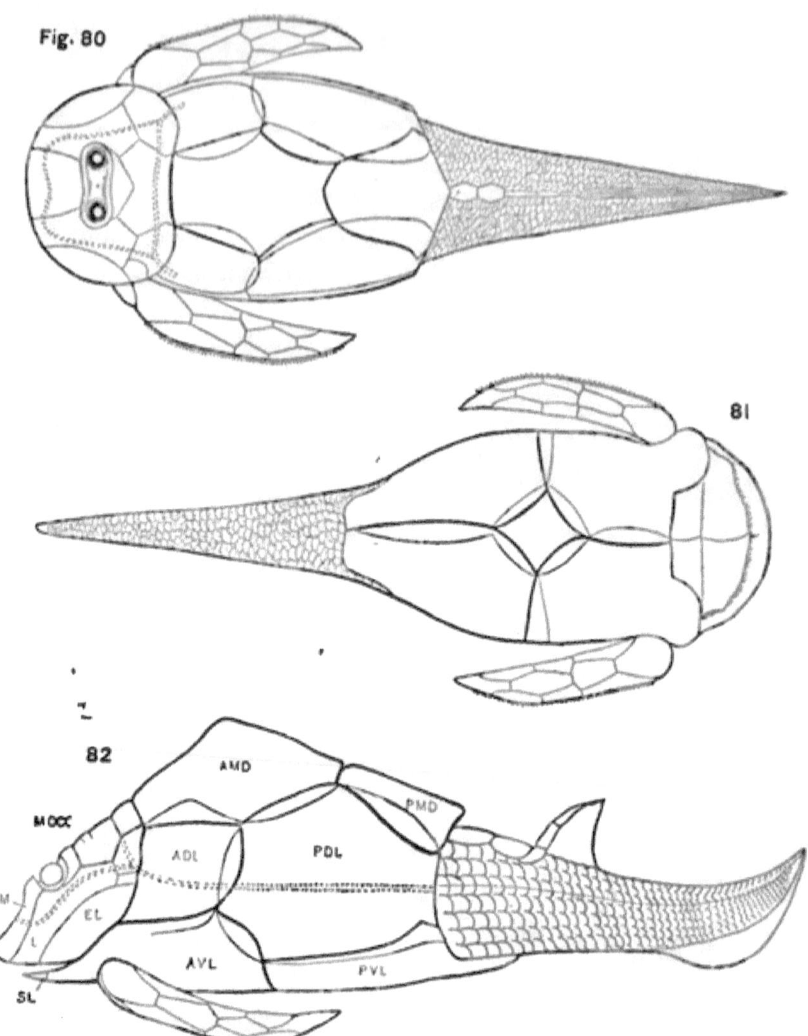

Figs. 80–82. — *Pterichthys testudinarius*, Ag.; restored by R. H. TRAQUAIR, from the dorsal aspect (80), ventral aspect (81), and lateral aspect (82). The double dotted lines indicate the grooves of the sensory canal system; and in the trunk, the thick lines represent the exposed borders of the plate, the thin line showing the extent of the overlap.

ADL. Anterior dorso-lateral. *AMD.* Anterior median dorsal. *AVL.* Anterior ventro-lateral. *EL.* Extra-lateral (or operculum). *L.* Labial. *MOCC.* Median occipital. *PM.* Premedian. *PDL.* Posterior dorso-lateral. *PMD.* Posterior median dorsal. *PVL.* Posterior ventro-lateral. *SL.* Semilunar. (Figure from SMITH WOODWARD.)

organs. The angles of the head plate are in some genera produced most acutely, and bear spines which served probably in progression. The body walls were encased in metameral derm plates, which became arched in the median line to serve as a dorsal fin. A heterocercal tail and an anal fin were also present. Problematical opercular flaps protruded at the sides of the head plate, and represented (as is now known) a continuation of the elastic middle layer of the head plate.

Pterichthys must be looked upon as the culminating type of these anomalous forms (Figs. 80–82). As in some Cephalaspids, there are two body regions that are cuirassed, — head and thorax. The tail portion is encased in dermal plates; it bears a dorsal fin and a clumsy heterocercal tail. In the consolidation of its armoured parts the elements are usually clearly indicated. The curious arm-like jointed appendages at the lateral head angle were formerly regarded as homologous with the opercular flaps of Cephalaspid, but are now known to be nothing more than the lateral head angles produced and specialized (*i.e.* jointed for locomotion). The strengthening spine of the dorsal fin is also but a primitive specialization of the body integument; it is formed by a pair of the bent scales of the dorsal ridge, and is not, therefore, homologous with the radial fin cartilages of fishes.

In Cephalaspids and Pterichthids there occurs a pineal plate (or its equivalent) which may have been either movable or fixed. In this are to be found the paired eyes and the socket of a median unpaired eye (?). In all of these singular forms mouth parts * are wanting. In

---

* Smith Woodward has since described a pair of inturned labial plates in the mouth of Pterichthys. Their position suggests that the sides of the mouth rim might become apposed, as in the Cyclostomes.

no instance has a trace of endoskeletal parts been observed.

The more that is determined of the structural characters of Ostracoderms, the less is it possible to accept the views as to their affinities with forms other than "fishes," either (Cope) as to their permanent larval-ascidian characters, or (Patten) as to their relationships with arachnids. Their general kinship is certainly to the fishes. According to Smith Woodward, the markings appearing on the visceral surface of head tests indicate the presence of gill pouches; in some forms clearly marked furrows suggest the possession of vertical semicircular canals; fish-like sense organs occur (Fig. 77); and their derm plates, in their cancellated and bone-like characters, cannot well be likened to the exoskeletal parts of invertebrates.

The lamprey-like form, *Palæospondylus gunni*, Traquair (Fig. 73), in the Lower Devonian is by many looked upon as the actual solution of the Cyclostome, and even of the Ostracoderm puzzle. This interesting fossil was discovered by Dr. Marcus Gunn, in the Lower Old Red Sandstone of Caithness, and was described in several papers by Traquair (*Trans. Edin. Soc.*, 1892–1894). It is of very small size, commonly of about an inch in length, but is admirably preserved (Fig. 73). There can be no doubt that Palæospondylus possessed a ring-like mouth surrounded by barbels like those of a Myxinoid, and that it lacked paired fins. But as a Cyclostome it must have highly specialized, having the same relation to the more primitive Cyclostomes of its day, as had the minute Acanthodians (p. 81) to the existing sharks. It had thus a remarkably large caudal fin with elaborately bifurcating supports; it had evolved stout, ring-like vertebræ, even in the caudal region, which had developed stout neural processes. Its skull was highly **evolved : in its anterior part** were represented, according **to** Traquair, the **palatine cartilages**; the brain case was complete, **and the auditory** capsules were of relatively **enormous** size. **The lateral plates** of the neck region are as yet uninterpretable.

From the evidence of Palæospondylus, accordingly, it may reasonably be inferred that lamprey-like forms existed in highly specialized conditions, even **at** the beginning of Devonian times. If they then existed, **it is** of course **not** impossible, **and** perhaps even not improbable, that **their** offshoots **may have** culminated in the **Ostracoderms,** as Smith **Woodward** has suggested. These can certainly belong **to** no gnathostome **stem. Their organs, though often** highly specialized, **were yet of the most primitive** order,—lack of paired appendages,[*] soft**ness of axial parts,** lowly sense organs; even the dermal plates, elaborate in their subdivision or ornamentation, or in the **special uses,** as "opercula," "pectoral **fins," or " fin rays,"**[†] are yet but primitive specializations **of the exoskeleton.**

---

[*] The presence of paired fins in Palæaspis, **as determined by Claypole, has** not been confirmed. The present writer, to **whom the type specimens were** kindly **shown by** their describer, must regard **these structures as elasmo-**branchian (Chimæroid?) spines, in crushed condition, **accidentally associated** with the head region of the fossil.

[†] It is obvious that these structures are but **analogous to the opercular and** fin structures of fishes, and would **tend to separate, rather than closen, the ties of kinship of these groups.**

IV

THE SHARKS

ALL true fishes may conveniently be grouped into the four sub-classes that have been noted (p. 8) in the introductory chapter. These are now in turn to be considered, and in this review the principal forms, fossil and recent, of each group must be exemplified. From the standpoint of their structural and developmental characters, a general idea of the mutual relationships of the fishes may finally be deduced.

The sub-class Elasmobranchii, which includes the sharks and rays, is usually regarded as representing most nearly the persistent ancestral condition of fishes, and, indeed, of all other jaw-bearing vertebrates. As a group it should certainly be taken first in the present discussion, as a convenient basis of comparison.

Sharks and rays should be looked upon at the beginning as the representatives of the oldest, most widely diffused, and possibly largest group of fishes. In their living forms they suggest but faintly the number and variety of their fossil kindred. It is generally thought that the history of this group, when more perfectly determined, is to furnish the most important evidence as to the general lines of descent of the fishes.

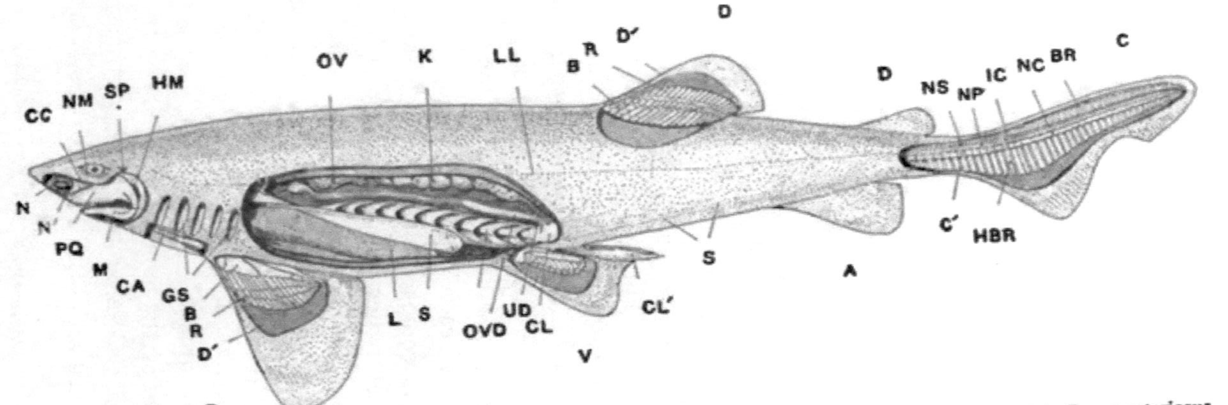

Fig. 83. — General anatomy of shark (♀).

A. Anal fin. *B.* Basal cartilaginous fin supports. *BR.* Basals and radials. *C.* Caudal fin. *C'.* Centrum. *CA.* Conus arteriosus, showing valves. *CC.* Cartilaginous cranium. *CL.* Cloaca, showing entrance of ureter, oviduct, and rectum. *CL'.* Clasping appendage of ventral fin. *D.* Dorsal fin. *D'.* Dermal fin rays. *GS.* Gill slits. *HBR.* Hæmal, basal, and radial supports. *HM.* Hyomandibular. *I.* Intestine, showing spiral valve. *IC.* Intercalary (interneural) plate. *K.* Kidney. *L.* Liver. *LL.* Lateral line. *M.* Mandible (Meckel's cartilage). *N.* Position of anterior and *N'* of posterior nasal aperture. *NC.* Notochord. *NM.* Nictitating membrane of eye. *NP.* Neural process. *NS.* Neural spine. *OV.* Ovary. *OVD.* Oviduct (Müllerian duct). *P.* Pectoral fin. *PQ.* Palatoquadrate cartilage. *R.* Radials. *S.* Stomach. *SP.* Spiracle. *UD.* Urinary duct. *V.* Ventral fin.

73

Structural Characters

The definition of a shark emphasizes its cartilaginous skeleton, investiture of shagreen, **uneven** (heterocercal) tail, and its **separate** and slit-like gill openings. **Its more** definite **characters may** well be summarized in **the** accompanying figure (Fig. 83).

I. **The skeleton is** cartilaginous (cf. **Fig.** 83, 84, and p. 252), **sometimes calcified** generally, **but** always (in recent **forms) lacking in dermal bones.** Behind the simple, trough-**like brain case the vertebral rod,** beginning at the occip-**ital condyles, is clearly segmented ;** the notochord is often **retained, especially in the tail region,** *NC,* but is encroached **upon by the cartilaginous rings, centra,** *C,* arising metamer-**ally in its sheath (Fig 85).** The vertical supports **of each centrum include a well-marked ventral plate, the** hæmal **arch and spine,** *HBR,*—which **in the tail region** probably **represents as well** the cartilaginous elements **of** the fin **support,—and a pair of small dorsal plates, the** neurals **and interneurals,** *NP, IC,* each **capped** by a neural spine, *NS.* **The fin supports compare closely in** structure **with the vertebral processes ; they form a** large part of the **functional fin, and** preserve **clearly, both in basal and radial parts, their metameral character.** This **segmental arrangement is also characteristic of** the supporting ele-**ments of the cavity of** the mouth and throat. These con-**stitute the "visceral arches"** (cf. p. **256)** which pass **backward from the rim of the mouth to** the region of the **pectoral fin.** The **first** visceral arch strengthens the rim **of the mouth ; it is margined with** teeth and functions as jaws,[*]

---

[*] The writer believes that the upper element of the mandibular arch **is to be regarded as** the palatoquadrate cartilage, rather than the pre-spiracular **ligament.**

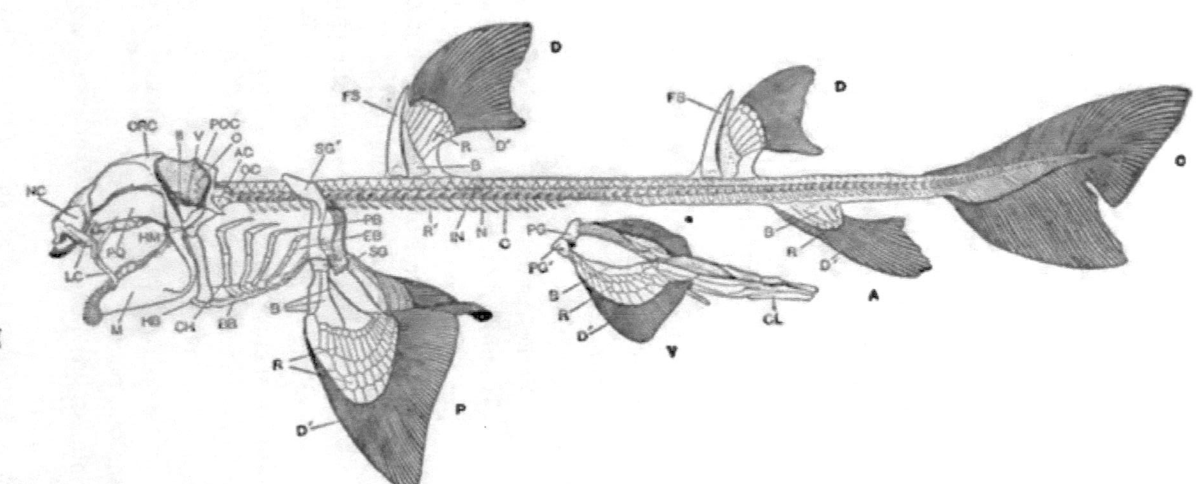

Fig. 84. —Skeleton of shark, *Cestracion galeatus* (♂). (From preparation of H. A. WARD. Drawing by ARNOLD GRAF).

A. Anal fin. *AC.* Auditory capsule. *B.* Basal cartilaginous fin supports. *BB.* Basibranchial. *C.* Caudal fin. *CH.* Ceratobranchial (the gill arch element immediately anterior is the ceratohyal). *CL.* Claspers. *D.* Dorsal fins. *D'.* Dermal rays of fin. *EB.* Epibranchial. *FS.* Fin spine. *HB.* Hypobranchial. *HM.* Hyomandibular. *IN.* Interneural plate. *N.* Neural process. *NC.* Nasal capsule. *O.* Orbit. *OC.* Occipital condyle. *ORC.* Orbital crest. *P.* Pectoral fin. *PB.* Pharyngobranchial. *PG.* Pelvic girdle. *PG'.* Dorsal process of *PG.* *POC.* Postorbital ridge. *PQ.* Palatoquadrate. *R.* Radial. *R'.* Rib. *SG.* Shoulder girdle. *SG'.* Dorsal process of *SG.* *V.* Ventral fin. *II.* Foramen of optic nerve. *V.* Foramen of fifth cranial nerve.

P and *M*.   The second arch serves as the principal support of the jaw hinge, *HM*, while holding in position, ventrally, the hinder arches; it also supports the tongue, and forms the hinder border of the spiracle (p. 19).   The succeeding arches, usually five in number, are the bearers of the functional gills, their jointed structure permitting the dilating and contracting movements of breathing.

As a further skeletal element of the Elasmobranchs the sub-notochordal rod is to be mentioned.   It is present in the larval stages of sharks, and appears to persist in the adult Cladoselache (p. 79).   It is a prominent structure of the hinder body region, passing along, like a second notochord, immediately below the vertebral axis. ˙Its significance is unknown.

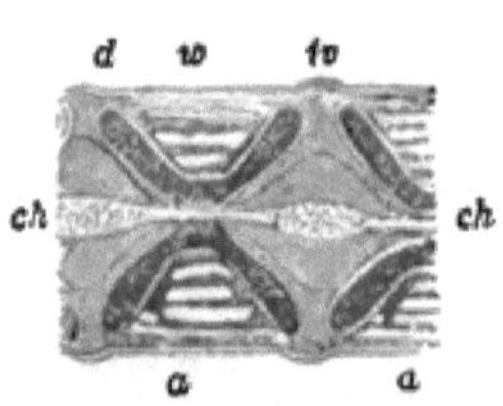

Fig. 85. — Vertebræ of shark (*Squatina*), longitudinal section. (After ZITTEL.)

ch. Notochord. *d*. Calcified rim and anterior surface of centrum. *iv*. Intervertebral space. *w*. Centrum.

II. The INTEGUMENT of the sharks, as has been noted (p. 23), is studded with shagreen denticles, often in metameral arrangement. These have been shown to correspond clearly with the teeth.

The soft structures characteristic of the Elasmobranchs include : —

III. GILLS, arranged metamerally (p. 19) ; the most anterior one partly functional in the spiracle, *SP*.

IV. SENSE ORGANS OF THE LATERAL LINE, in some forms in an open sensory groove, in others sunken and constricted in metameral pouches.

V. BRAIN, simple in its segmental characters and cranial nerves (v. p. 274).

VI. NASAL ORGAN, EYE AND EAR, as shown on p. 276.

VII. RENAL AND REPRODUCTIVE SYSTEMS (p. 270), abdominal pores (p. 271).

VIII. Digestive tube with a single bend, *S, I*, the intestine provided with a spiral valve (p. 263), **terminating**, together with the ducts of the renal and reproductive organs, in **a** common cloaca, *CL* (p. 266). **Liver, *L*,** spleen, and pancreas large ; mesenteries simple but greatly fenestrated ; **air** bladder absent.

IX. Heart with a contractile arterial cone, *CA,* containing several rows of valves (p. 260) ; circulatory system in general as described on **p.** 269.

X. "Claspers" developed at the hinder margin **of** the ventral fins as the intromittent **organ of the male.** They are rudimentary in the female, *CL'*. **Each clasper** is **the** trough-like **hinder rim of the fin, which becomes** transformed into the **compact,** elongated, tube-like sperm canal. Its tip is **often studded** with elongated shagreen denticles whose recurved cusps retain it *in copulo.*

Fossil Sharks

Of all fishes, sharks **certainly suggest most closely in** their general structures the metameral conditions of **the** Cyclostome: it should also be noted **that** they possess the greatest number **of** body **segments, in** some **instances** over three hundred, known among **vertebrates. Little** is known, however, of the primitive **stem of** the sharks, and even the lines of descent of the different members **of the** group can only be generally suggested. The development of the recent forms has yielded few results of undoubted value to the phylogenist : it would appear as if palæontology alone could solve the puzzles of their descent.

The history of fossil sharks has as **yet been** but imperfectly outlined. **The remains of the** more ancient forms have usually proven so imperfectly preserved that little could be determined of their structural characters. **Spines,** teeth, shagreen denticles, have proven the antiquity of the shark stem and the wealth and variety of its fossil forms; they have provided the evidence that even in Silurian times there lived sharks whose exoskeletal specializations had progressed further than in their recent kindred: that in the Carbon there occurred the culminating-point in their differentiation, when specialized sharks existed whose varied structures are paralleled only by those of existing bony fishes, — sharks fitted to the most special environment; some minute and delicate; others enormous, heavy, and sluggish, with stout head and fin spines, and elaborate types of dentition.

But the detached fragments of the fossil sharks can give little satisfactory knowledge of their general structures. The simpler the form of the shark, indeed, the less liable is it to become fossilized. The more generalized of the ancient sharks must thus remain structurally unknown until more perfect fossils come to be found. To this event the discoveries of the past few years have certainly yielded most encouraging aid. Several forms of sharks of the Lower Carbon and Permian have been obtained in a condition of admirable preservation, and have already contributed materially to the morphology of Elasmobranchs. Other early forms may be forthcoming which will be found to have retained sufficient of the characters of their ancestors to warrant more definite views as to the general relationships of fishes.

Of the three primitive forms of fossil sharks lately described: the earliest, from the Ohio Waverly (Lower Carbon) is *Cladoselache*, Dean; a later and puzzling form, from the Carboniferous, is *Chondrenchelys*, Traquair; the latest from the Permian and Coal Measures, is *Pleuracanthus*, Agassiz. The only early shark type that had previously been structurally known was that of the aberrant and highly specialized *Acanthodian* of the Coal Measures.

Cladoselache is the most primitive, as well as the oldest, of these ancient sharks. It is relatively of small size, varying in the length of its species from two to six feet. Its outward form, as restored by the writer, is seen in Fig. 86, and in ventral view in Fig. 86 *A*. The shape is clearly

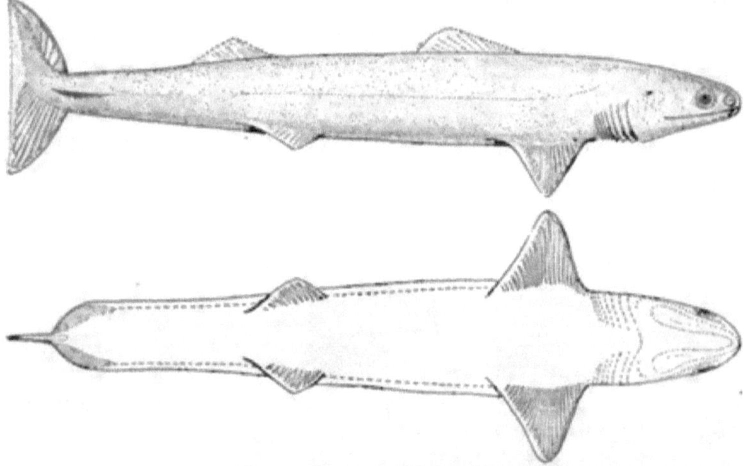

Fig. 86. — *Cladoselache fyleri*, Newb. × ¼. Restoration by writer. After specimen in the museum of Columbia College from Cleveland shales, Ohio.
Fig. 86 A. — *Cladoselache fyleri ;* ventral aspect.

that of a modern shark; the fins, too, in their size and position, have somewhat of a modern look; and at the base of the tail occurs the small horizontal keel of many living forms. But in spite of these peculiarities, Cladoselache must be looked upon as the most archaic, and, in many ways, the most generalized of known sharks; its paired fins are but the remnants of the lateral fold (p. 43), serving alone as balancers; the tail, curiously specialized, is widely heterocercal, its hinder web lacking supports in the upper lobe (p. 36); the vertebral axis is notochordal;

and the writer now finds that an exceedingly simple condition existed in the neural and hæmal arches ; they prove to be of moderate size and thickness, each a tapering rod of cartilage, forked at its base ; each body segment contains a single neural and hæmal spine, closely alike in size. Unlike modern sharks, Cladoselache was without claspers : its eggs must have been fertilized after their deposition, as in the majority of fishes other than Elasmosbranchs. The gill openings, at least seven (probably nine) in number, appear as in the restoration, to have been shielded anteriorly by a dilated dermal flap. A spiracle was probably present. The jaws were slender, and apparently hyostylic (p. 257) ; [*] the teeth are of the pattern of shagreen denticles, but occur

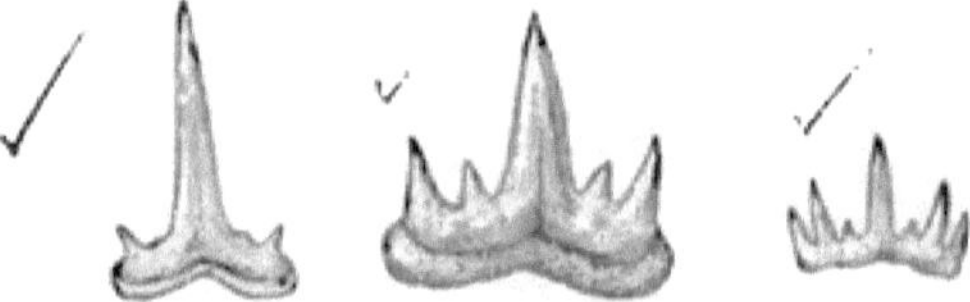

Fig. 86 B. — Teeth of ("Cladodus") Cladoselache. × ⅓. The above forms occur in different regions of the mouth.

in clusters ("*Cladodus*," Fig. 86, B). The mouth was terminal in its position. The nasal capsule was apparently not connected with the mouth by a dermal flap. The eye was protected by several rings of rectangular plates, clearly shagreen-like in character. The integument was finely studded with minute lozenge-shaped denticles, and was everywhere lacking in membrane bones. The lateral line retained its groove-like character.

The shark, Acanthodes (Fig. 87), of the Coal Measures is now to be regarded (Smith Woodward) as a member of a highly specialized Palæozoic group. And its many specialized structures — added to its greatly reduced size —

---

[*] As Claypole's recent figure seems to demonstrate. *Am. Geol., Jan. 1895.*

may, perhaps, have been the cause of its extinction.
The present writer believes that Cladoselache may well
have represented the ancestral form of the Acanthodian.
The generalized structures of the former have given place
to a perfected dermal armouring, and a completed series

Fig. 87. — *Acanthodes wardi*, Egert. × about ½. (Restoration slightly modified after SMITH WOODWARD.)   Coal Measures, England.

of balancing fins.   In Acanthodes the shagreen denticles
have thus become greatly enlarged and thickened, their
flattened and enamelled surfaces wedging closely together (Fig. 88); and on the roof of the head and
mouth traces of membrane bones have appeared.   Around

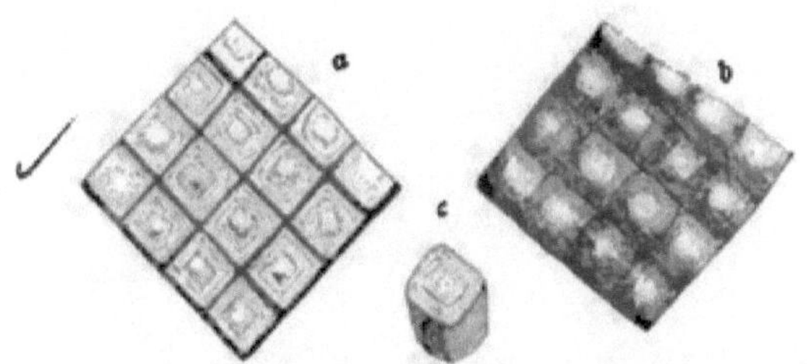

Fig. 88. — *Acanthodes gracilis*, Beyr.   Shagreen.   × about ¹⁸⁰.   (After ZITTEL.)
a. Outer face.   b. Inner face.   c. Isolated denticle.

the eyes the many shagreen plates of Cladoselache have
fused into a group of four.   Supporting the dermal gill
frills, there have also appeared rows of minute sculptured
plates (corresponding, perhaps, to those, *BR*, of Fig.
145), homologous, apparently, with shagreen denticles.

G

Further resemblances to Cladoselache are to be traced in
the position of the fins, gill slits, eyes, mouth, nasal capsule, and in the structures of the caudal fin (Kner), and of
the lateral line.   The teeth, however, are no longer of the
derm-denticle pattern ; they have become few in number,
large, and "degenerate" in their fibrous structure (Fig.
88, *A*).   The fins are clearly more perfect balancing organs than those of
the older shark ; their anterior rim is
formed by a stout spine, representing,
the present writer believes, the concrescence of the radial fin supports ;
it is heavily crusted over with the
calcifications of shagreen denticles.   The functional fin
area has thus become dermal, and is lacking in supports,
excepting in the pectoral fin.   This, as the most highly
specialized of all the body fins (p. 41), appears in some
cases to have evolved 'strengthening (dermal) rays in its
proximal portion (as in Figs. 87 and 32).

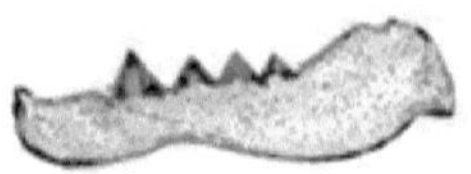

Fig. 88 A. — Teeth of
Acanthodopsis wardi. ✕1.
From sketch after speci-
men in British Museum.

Fig. 89. — *Climatius scutiger,* Egert.   ✕1.   (From ZITTEL, after POWRIE.)
Old Red Sandstone, Forfarshire.

In connection with these fin structures the remarkable
Acanthodian, *Climatius* (Fig. 89), should finally be mentioned.   In this form the paired fins are represented by a
series of fin spines whose size grades backward from the
pectoral region ; a series of paired fins appear, therefore,

to have been present, and suggest strongly continuous fin-fold characters. (V. p. 40.)

Pleuracanthus (Fig. 90), the third of the well-known Palæozoic sharks, is widely different from the Acanthodian: it suggests a transitional form between the generalized Cladoselachian, on the one hand, and the Dipnoan on the other; or, more accurately, it demonstrates that the stems of shark and lung-fish were at one time drawn very closely together. It has thus far occurred only in the Carbon and Permian, but may reasonably be expected in lower horizons as a contemporary of the earliest lung-fishes.

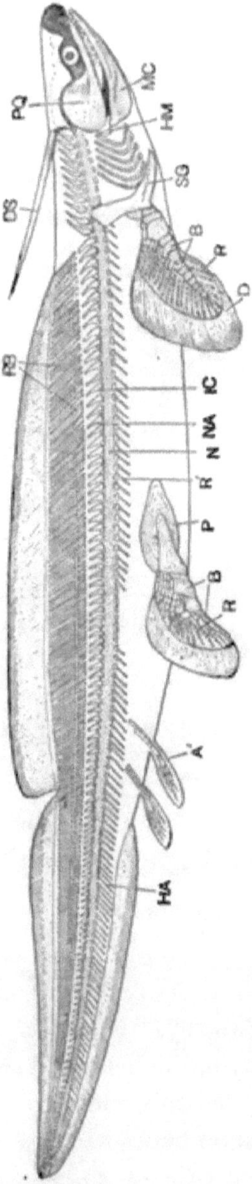

Pleuracanthus is in many ways the most interesting and suggestive member of the shark group; for it destroys many of our conventional ideas as to the general characters

Fig. 90. — *Pleuracanthus decheni* (Goldf.), ♀. × about ⅓. (Restoration slightly modified after A. FRITSCH.) From the Permian of Bohemia.
A'. Anal fin. *B.* Basal fin cartilages. *D.* Dermal margin of fin. *DS.* Dermal head spine. *HA.* Hæmal arches. *HM.* Hyomandibular. *IC.* Interneural plates. *MC.* Mandible (Meckel's cartilage). *N.* Notochord. *NA.* Neural process and spine. *P.* Pelvic cartilage (girdle). *PQ.* Palatoquadrate. *R.* Radial fin cartilages. *R'.* Rib. *SG.* Shoulder girdle.

essential to sharks. That it was actually a shark cannot
be doubted; its gills, six or seven in number, opened
separately to the surface; its teeth (Fig. 90 *A*) were
typically shark-like, arranged in many rows on Meckelian
and palatoquadrate cartilages; a tuberculated dorsal spine
was present; claspers occurred in the male; the vertebral
column, although notochordal, *N*, presented intercalary

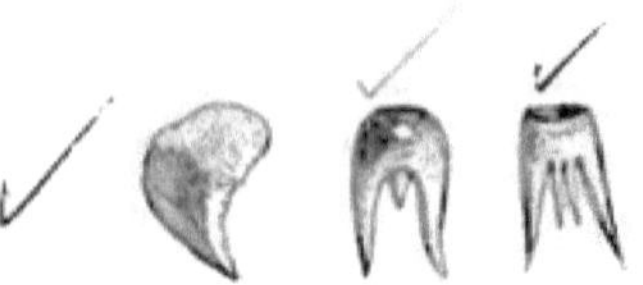

Fig. 90 A. — Teeth of *Pleuracanthus.* × ⅓.
(After DAVIS.)

plates, *IC*, and the
jaw was essentially
hyostylic, *HM*. On
the other hand,
many of its struct-
ures are clearly tran-
sitional to the Dip-
noan : the pelvic fins
are shark-like, with
the radial supports,
R, arising from but
one side of the line
of basals, *B* ; but the
pectoral fin is typi-
cally archipterygial,
and the caudal diphy-
cercal, as in the lung-
fishes. In this re-
gard the continuous

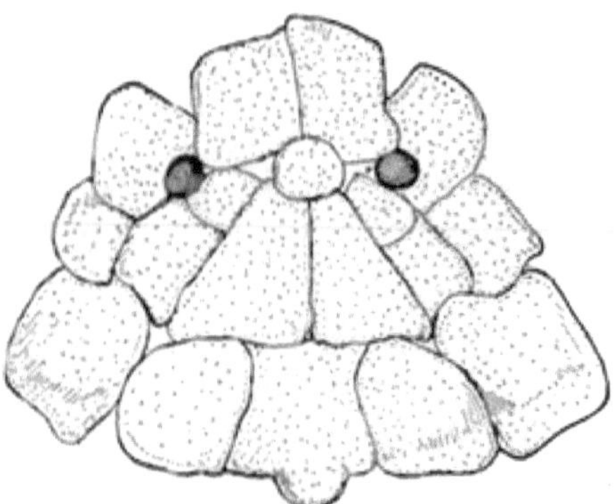

Fig. 90 B. — Dermal bones of the head roof of
Pleuracanthus. × ⅓. (After DAVIS.)

dorsal fin, with its separate basals and radials, *B* and *R*, is
again noteworthy. But most singular of all the features
of this lung-fish-like shark were its integumentary characters ; shagreen tubercles had disappeared on the body surface, and derm bones had appeared roofing the head : their
arrangement (Fig. 90 *B*) is strikingly similar to that of the
lung-fish of Fig. 124.

The final form of Palæozoic shark whose structural characters have in any way been described is Chondrenchelys. It appears to have somewhat resembled the Pleuracanthid in its elongate form and tapering tail; but as yet the details of its structure have not been discovered. In its vertebral characters it had certainly made a marked advance; the notochord had become greatly constricted; and well-marked centra and arches were present. These appear to have been highly calcified, and show a peculiar

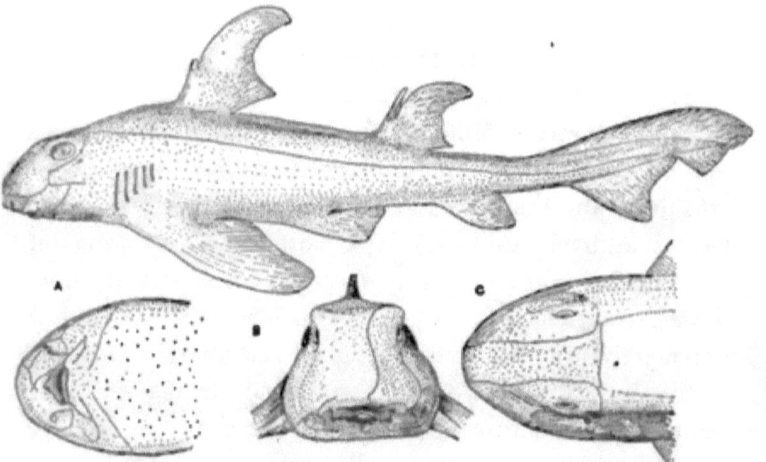

Fig. 91. — Port Jackson shark, *Cestracion philippi* (♀). × ⅒. (After GARMAN.) Australia. *A*. Ventral. *B*. Anterior, and *C*. Dorsal aspect of head.

beaded or fretted structure which in this form is apparently unique.

Other ancient sharks, as far as can be inferred from fragmental structures, appear to have closely resembled forms that are still extant.

Such unquestionably were the Cestracionts, a group of sharks especially abundant in the early Palæozoic seas, judging from the numbers of their fin spines and pavement teeth that have been preserved. Their bygone rôle was certainly a long and important one. In some of their forms they could have differed but little from their single survivor, the Port Jackson shark, *Cestracion* (Heterodontus) (Fig. 91, *A*, *B*, *C*). In others, the dentition and dermal defences suggest a wide range in evolution. Their general character appears to have been primitive, but in structural details they were certainly specialized ; thus their dentition had become adapted to a shellfish diet, and they had evolved defensive spines at the fin margins, sometimes even at the sides of the head. In some cases the teeth remain as primitive shagreen cusps on the rim of the mouth, but become heavy and blunted behind ; in other forms the fusion of tooth clusters may present the widest range in their adaptations for crushing ; and the curves and twistings of the tritoral surfaces may have resulted in the most specialized forms of dentition (*e.g. Janassas, Petalodonts, Cochliodonts*, and *Psammodonts* of the Coal Measures) which are known to occur not merely in sharks but among all vertebrates. Equally interesting may prove the evolutional details of other cestraciont structures when they come to be known. Ray-like proportions may well have been evolved even among the earliest Palæozoic forms.

The surviving member of this group, Cestracion, suggests in itself the adaptations of a bottom-living form in its greatly enlarged pectorals. Its genus, however, has not been traced earlier than the Mesozoic, although its comparatively generalized dentition (Fig. 27) suggests a far more remote descent.

It is of interest to note that Cladoselache approaches in its dentition the characters of the primitive Cestracionts (*e.g. Synecchodus*).

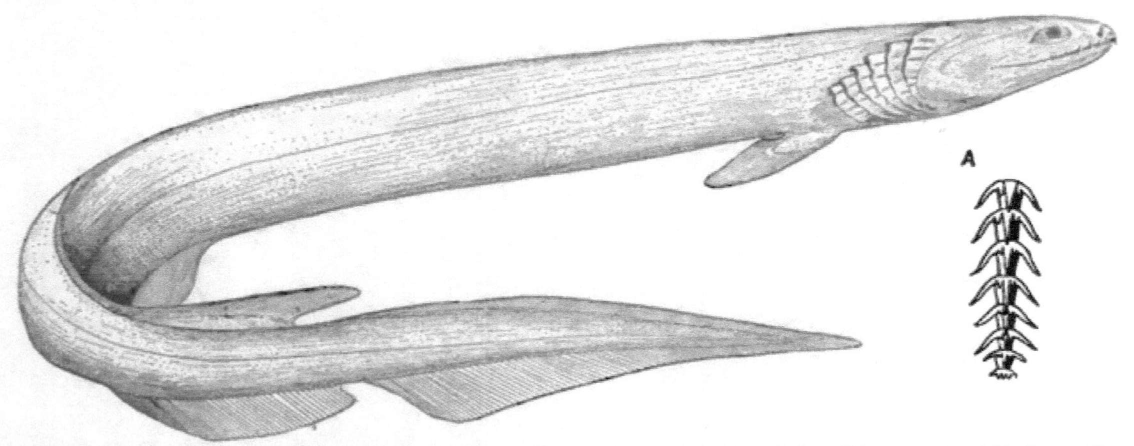

Fig. 92.—Frilled shark, *Chlamydoselache anguineus*, Garm. ♀. × about ⅛. (After GÜNTHER, in "Challenger.") Madeira Islands and Japan.

A. Row of teeth (ecto-entad), enlarged.

Recent Sharks

The forms of Sharks and Rays common at the present time are generally looked upon as closely related genetically, although their lineage cannot be definitely traced. As far as palæontological evidence goes, they may well have been derived from a single Palæozoic ancestor.

Perhaps of all recent forms, *Chlamydoselache* (Fig. 92), and *Notidanus* (Heptanchus, or Heptabranchias) (Fig. 93), which are universally regarded as "primitive," have inherited most directly the features of this generalized Palæozoic form. But which of these two sharks must be regarded as resembling its remote ancestor the more closely seems to the writer a very doubtful matter. Chlamydoselache derives its great interest from its late discovery (1884, Garman), rareness, and Pleuracanthid type of teeth (Fig. 92, *A*); but now that it has been taken in numbers — comparatively — in deep water, one is inclined to believe that many of its

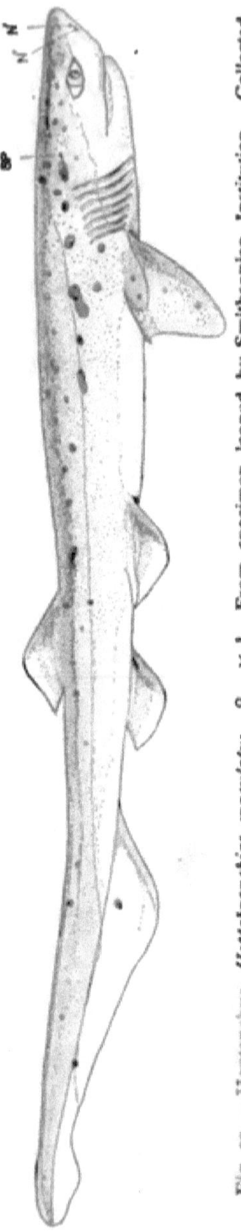

Fig. 93. — Heptanchus, *Heptabranchias maculatus.* ♀. × ¼. From specimen loaned by Smithsonian Institution. Collected in Pacific.
N. N'. Anterior and posterior nares. *SP.* Spiracle.

"primitive" features, like its eel-like shape, may partly be due to its environment : its resemblance, moreover, to the Pleuracanth has since been found to be of a superficial character. Notidanus, on the other hand, adds to its primitive characters the presence of no less than seven

Fig. 94. — The horned dog-fish, *Squalus acanthias*, L. ♂. × ⅓. (After GOODE· in U. S. F. C.) Atlantic.

gill slits, — a feature which morphologists generally are inclined to regard as of great significance.

The many forms of recent sharks have certainly not diverged widely from the stem of descent which Notidanus may well represent : they retain the sub-cylindrical body form, specializing more or less to environment; in deep-sea genera the body length appears proportionally in-

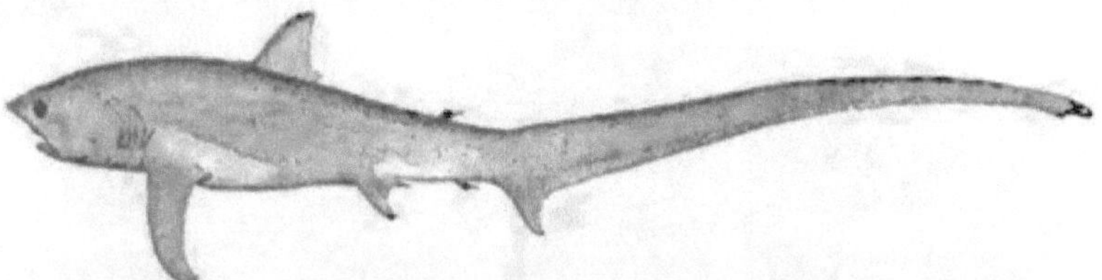

Fig. 95. — The thrasher shark, *Alopias vulpes* (Gmel.), Bonap. ♀. × ⅟₁₀. Atlantic.

creased : predatory forms, such as *Squalus, Alopias, Lamna* (Figs. 94, 95, 96), acquire great size and strength, travel great distances, and are enabled to become cosmopolitan. Among the minor details to which their evolution has been carried, may be noted : the pattern, size, and arrange-

ment of teeth and shagreen denticles; the calcification of
the vertebræ (great differences sometimes occurring in the
same genus, *e.g. Scyllium*), the size, disposition, and num-

Fig. 96. — The mackerel shark, *Lamna cornubica* (Gmel.), Fleming. × ⅙.
North Atlantic.

ber of the fins, the more or less pouch-like character of
the sensory canals.

In the basking shark, *Cetorhinus* (Selache) (Fig. 96 *A*),
widely specialized conditions occur in the gill rakers,
which enable the throat to retain the smallest food organ-

Fig. 96 A. — The basking shark, *Cetorhinus maximus*, (L.) Blainville. ♂.
× ⅛. (After Goode in U. S. F. C.)

isms.   In another shark, *Læmargus* (Fig. 96 *B*), the eggs
are probably fertilized after being deposited, — a condition
unique among recent Elasmobranchs.

The different families of the existing sharks appear to
to have been already differentiated during the early Mesozoic times. The ancient shark-like form had then given
place to the flattened and rostrated types, adapted to the

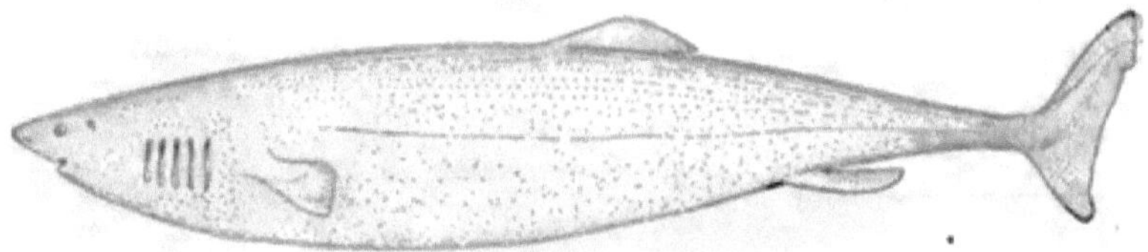

Fig. 96 B. — The Greenland shark, *Læmargus borealis*, L. × ᵢ̅. (After GÜNTHER.)

conditions of bottom living and to the special character of
their shell-fish or crustacean diet.

One of the earliest offshoots from the main selachian
stem appears to have been *Squatina* (Fig. 97), popularly
known as the monk-fish, or angel-fish. As early as the
Mesozoic times it was existing, differing but little from the
recent species. Its general shape is shark-like, although its

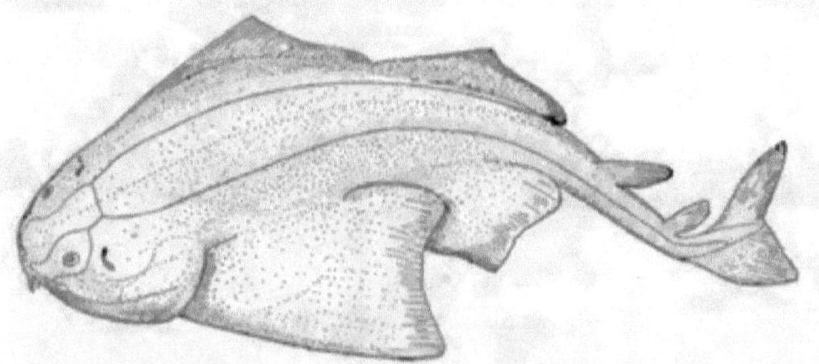

Fig. 97. — The monk-, or angel-fish, *Rhina squatina*. ♀. × ᵢ̅. Atlantic,
Mediterranean, Pacific.

head and trunk are clearly depressed. This, together with
its enlarged pectoral fins, enables it to take a position
closer to the bottom.

The recent saw-fish, *Pristis* (Figs. 98, 98 *A*), is next to

be mentioned as a form somewhat transitional from shark
to ray.   Its body, as may be seen in the figure, has been
strikingly flattened, the gill openings changing their position from the lateral to the ventral side, but the fins retaining in general the selachian characters.   Its singular
rostrum with lateral spike-like teeth is unquestionably a

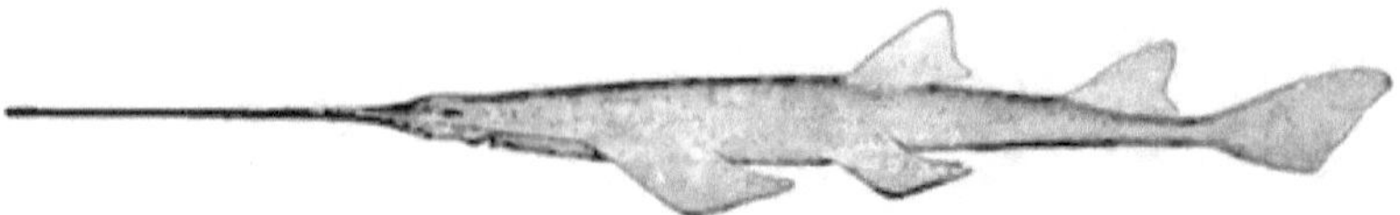

Fig. 98.—The saw-fish. *Pristis pectinatus*, Latham. ♀. × ⅒. Tropical
seas. (After GOODE in U. S. F. C.)

highly specialized organ.   Pristis is thus far known not
earlier than the Eocene, but its close connection genetically
with the ancient and more generalized *Pristiophorus* is
usually conceded.

Pristiophorus (Fig. '99) is certainly more closely allied
to the sharks : its gill slits have not as yet acquired their
ventral position, and its rostrum suggests the ancestral

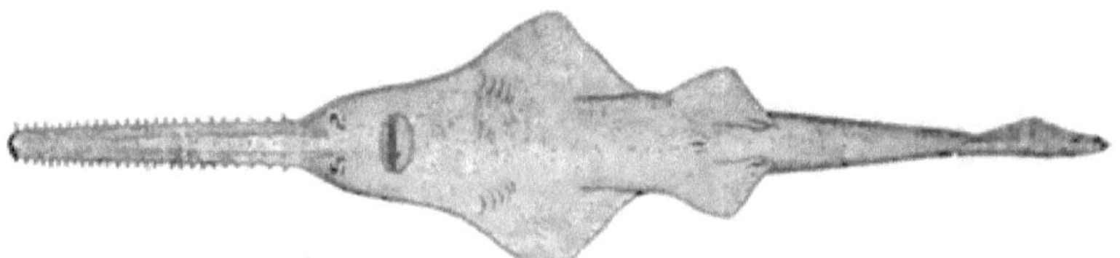

Fig. 98 A.—Saw-fish, ventral view.

conditions of that of Pristis.   Its barbel-like structures,
however, distinguish this form clearly from all other
Elasmobranchs.   It is known to have occurred as early
as the Jura.

The *Skates* or *Rays* are well known to represent the
most highly modified survivors of the ancient stem of the

sharks ; they appear comparatively late in time, and may well be regarded as the culminating forms of the specializing bottom-living sharks of the Mesozoic. Whether they are directly descended from forms like Squatina or Pristio-

Fig. 99. — *Pristiophorus (cirratus)*. ♀. (After JAEKEL.) Australia.

phorus must be looked upon as exceedingly doubtful, as the depressed body form may possibly have arisen independently in these different families. The most nearly ancestral form of the skates appears to have sur-survived in *Rhinobatus* (Fig. 100). The shark-like body form is here most nearly retained, and its fin structures

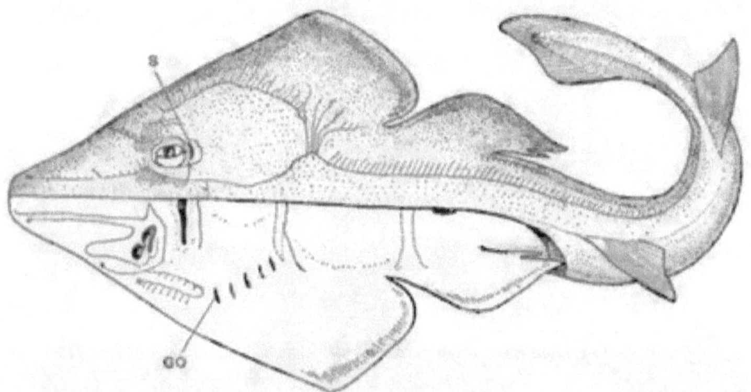

Fig. 100. — *Rhinobatus planiceps*. ♀. × ¼. (After GARMAN.) (The lower portion of the figure showing ventral side.) *S*. Spiracle. *GO*. Gill slits.

are the least specialized ; these transitional characters of Rhinobatus become more prominent in view of its ancient occurrence : its genus was clearly defined as early as the Oölite.

The body form of the Skate (Fig. 101) has become admirably adapted to bottom living; it is exceedingly flattened anteriorly, its head and trunk and paired fins fusing so perfectly that from the surface view one could not define their limits; the tail region, on the other hand, has dwindled away to rod-like or whip-like proportions.

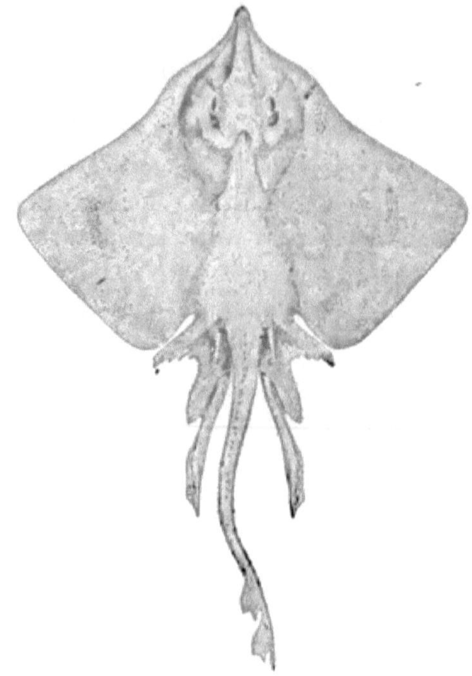

Fig. 101.— The barn-door skate, *Raja lævis*, Mitch. ♂. × ⅛. (After GOODE in U. S. F. C.)

In the process of flattening, the gill openings take their appearance early in the ventral side of the body, and the pectoral fins, enlarging rapidly, press closely forward at the side of the flattened head, fusing with its tissues. Motion is now accomplished by the gentle undulation of the long horizontal fin margin: and the enlarged anterior element of the fin stem, by being raised or depressed, comes to direct the upward or downward motion of the fish. In this mode of movement seems to have been paralleled the undulation of the ancestral fin fold. On the fish's dorsal side colour adaptations have become marked, the ventral aspect becoming deficient or wanting in pigment. In its habits the skate mimics the colour of the bottom and glides along inconspicuously, apparently without movement; when alarmed, it will press its enlarged and flattened fins so closely to the bottom that it appears to adhere, and is to be dislodged only with the greatest efforts.

Two of the aberrant forms of rays are shown in Figs. 102 and 102 *A*. The for-

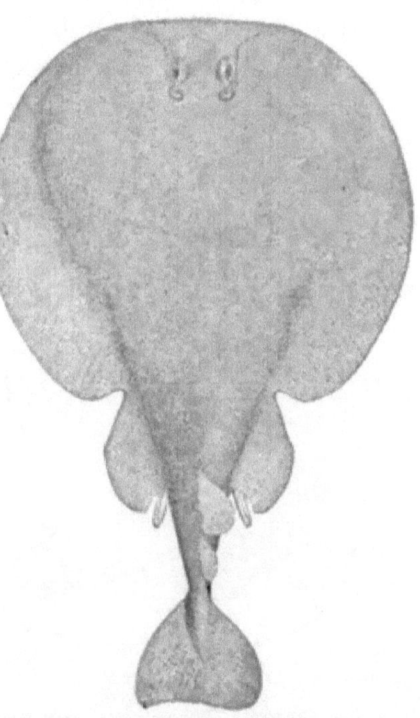

Fig. 102. — The torpedo, *Torpedo occidentalis*, Storer. ♂. × ¼. (After GOODE in U. S. F. C.)

mer, the *Torpedo*, is remarkable on account of its electric organs; the latter, *Dicerobatis*, on account of the great breadth of its pectorals, and its enormous size.

Affinities

In concluding the present chapter, the probable affinities and interrelationships of the Elasmobranchs may be summarized as follows (v. Fig. 103) :—

1. Of all known stems that of the shark is most nearly ancestral in the line of jaw-bearing vertebrates.

2. A generalized form not unlike Cladoselache might well represent the ancestor of Pleuracanthid, as well as of the primitive Cestraciont, of Acanthodes, and of the modern sharks and rays.

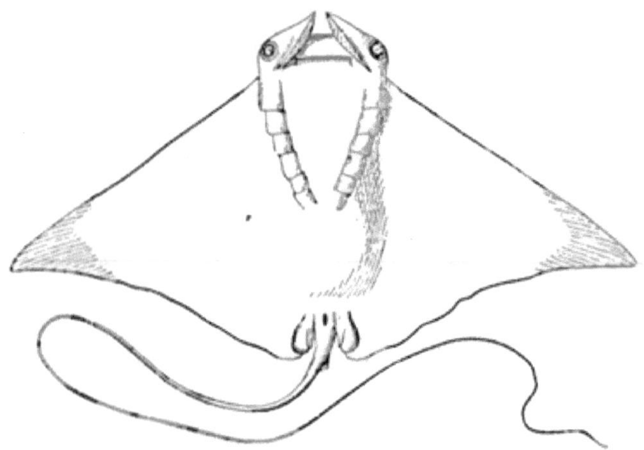

Fig. 102 A. — The mantis, or devil ray, *Cephaloptera (Dicerobatis) draco.* × ⅕. (After GÜNTHER.)   Tropical seas.

3. On the evidence of the Permian Pleuracanthids, lung-fishes (Dipnoans) and the earliest bony fishes (Crossopterygians) are to be derived from an advancing shark type.

4. From the ancestral stem of the recent sharks Cestracionts were the most early differentiated : it is one of their more generalized forms, Cestracion, that has alone

survived among the widely evolved **genera and families of** Palæozoic **times.**

5. The more primitive types of modern sharks, **Chlamy-**doselache, Notidanus, represent in **an almost** differentiated condition the Palæozoic phylum.

6. The modern rays are **derived** in early Mesozoic **times** from the main shark **stem, not (in** the opinion of the writer) descended from Cestracionts, Pristids, Pristiophorids, or **even (?)** Rhinids.

7. Chimæroids, next to be discussed, represent the most ancient of known offshoots from the (Pre-Silurian ?) sharks : they are not degenerate **in their essential** structures, nor are they connected **with the ancestral phylum of** the lung-fishes, save through **a common descent from early shark-**like **ancestors.**

These results the writer **has expressed in the** diagram on the following page. **The diverging phyla** are indicated **as** they are represented **historically ; their** primitive concurrence with the **main line of descent is** suggested **by** dotted lines.

H

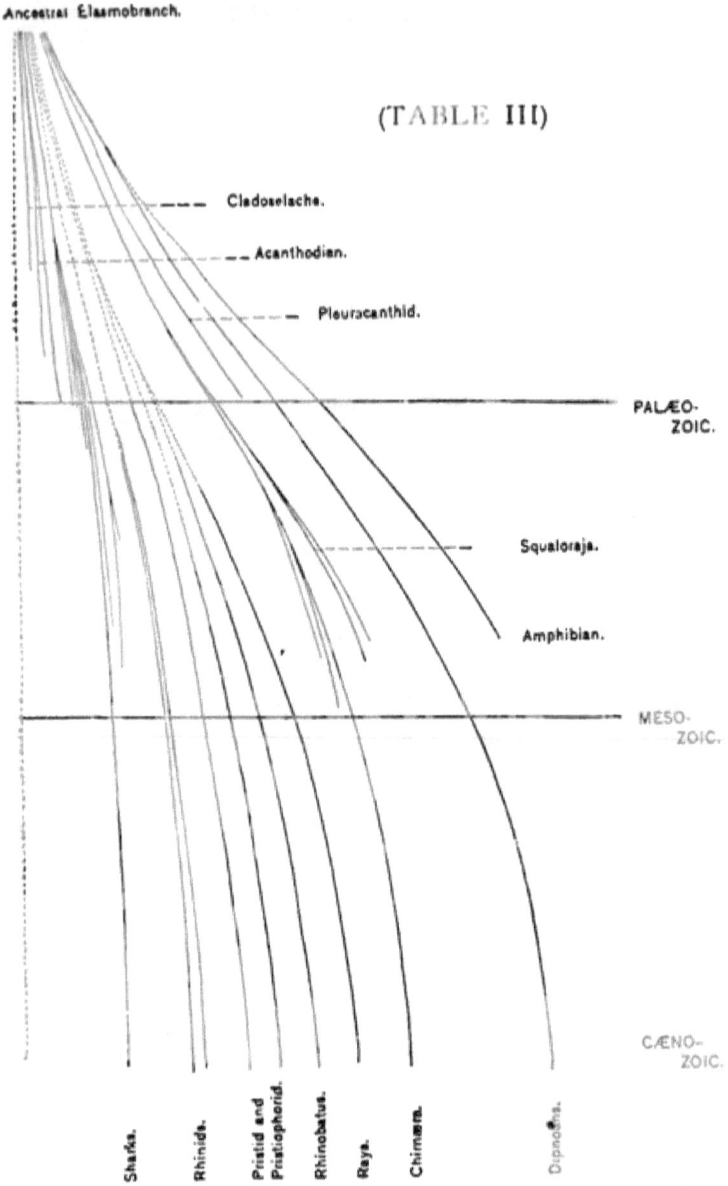

Fig. 103. — Scheme suggesting the interrelationships of Sharks, Chimæroids, and Lung-fishes.

98

V

THE CHIMÆROIDS

CHIMÆROIDS are shark-like in their general characters, but cannot be looked upon as in any **strict** sense closely associated with the Elasmobranchs. They constitute the second of the more important groups of fishes. **Their** typical representative is the Chimæra, spook-fish, **or sea-cat** (Fig. 119).

Structural Characters

The typical **structures of** Chimæra are shown **in the** dissection given in **Fig.** 104. **Its** thick, round, and **blunted** head tapers away gradually **to the tip of a** diphycercal **tail,** *C.* **The body** surface **is** generally smooth. The paired fins are somewhat shark-like, but their **dermal** margins have become greatly enlarged, tapering distally to an acute **point ;** the foremost dorsal fin provided **with** an anterior spine **folds** like a fan and may **be** depressed into a sheath, *SH,* **in the** body wall; this fin and the hinder ones are largely dermal, *D',* basal and radial supports existing only at *B', R'.* **The** gill arches, *BA,* may be seen to be closely drawn **together ;** their outer openings **are now** reduced **to the** slit-like aperture beneath **the** dermal flap, *OP.* Teeth **exist in the** form of dental plates, **closely** fused with **the jaws ; as** shown in the figure, *D,* three of these occur **in** each side, a single one on the mandible, an anterior **and** posterior on

99

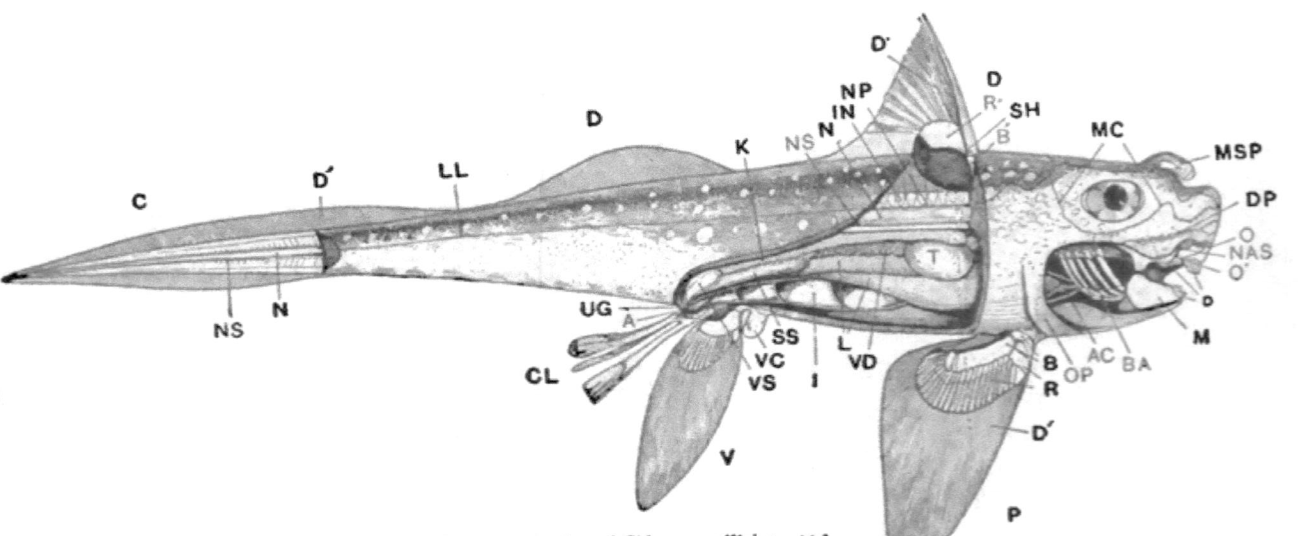

Fig. 104. — General anatomy of Chimærold. From preparation of *Chimæra colliei*, ♂. × ⅑.
A. Anus. *AC.* Arterial cone. *B.* Basal cartilaginous fin supports. *B'.* Fused basals of anterior dorsal fin. *BA.* Branchial arches. *C.* Caudal fin. *CL.* Claspers. *D.* Dorsal fin. *D.* Dental plates. *D'.* Dermal portion of fin. *DP.* Dermal plates of sub-orbital mucous groove. *I.* Intestine, showing spiral valve. *IN.* Interneural plates. *K.* Kidney (mesonephros). *L.* Liver. *LL.* Lateral line. *M.* Meckel's cartilage, articulating directly with cartilage of cranium. *MC.* Mucous canals. *MSP.* Frontal spine of male. *N.* Notochord. *NAS.* Nasal capsule. *NP.* Neural process. *NS.* Notochordal sheath. *O, O'.* Anterior and posterior nasal openings. *OP.* Dermal flap functioning as operculum. *P.* Pectoral fin. *R.* Radial. *R'.* Fused radials. *SH.* Sheath of anterior dorsal fin. *SS.* Sperm sac. *T.* Testis. *UG.* Urinogenital opening. *V.* Ventral fin. *VC.* Claspers of ventral fin. *VD.* Vas deferens (Wolffian duct). *VS.* Sperm vesicle.

the upper jaw ("premaxillary" and "palatine"); they are studded with hardened points, or "tritors" (Figs. 109–112). The sense organs are similar to those of sharks; the nasal capsule, *NAS*, has both an anterior and a posterior opening, *O, O'*, the latter within the cavity of the mouth.

The visceral parts are decidedly shark-like; the digestive tube is straight (p. 263); the intestine, *I*, with a spiral valve of three turns; the liver, *L*, is prominent; the kidney, *K*, reproductive organs, *T*, and their ducts, *VD, SS, VS*, and abdominal pores are as in sharks; the intestine, however, opens directly to the surface, *A*, separating an anal from a urogenital aperture, *UG*. The mesenteries are string-like.

The male fish is provided with a highly specialized intromittent organ, *CL*; it has a supplemental clasping organ, *VC*, at the front margin of each ventral fin, *V* (cf. also Fig. 116 and Fig. 116*a*), and a retractile spine in the region of the forehead, *MSP* (cf. Figs. 113 and 115).

The skeleton of a Chimæroid is shown in the following figure (Fig. 105). Its structure is cartilaginous. The vertebral axis is notochordal; its sheath, lacking in definite centra, is strengthened anteriorly by a series of calcified rings. In the anterior region of the trunk, neural processes, interneurals, and neural spines, *NP, IN, NS*, together with hæmal processes, occur as in sharks; toward the tail region they fade away, and before joining with the head at the occipital condyles, *OC*, they fuse into a compact mass, joining with the basal supports of the dorsal fin.

The cranium is of a highly compacted structure; its vertical height has been greatly produced; the orbits, *OR*, are of great size and are separated from each other by a membranous septum. The snout region is greatly meta-

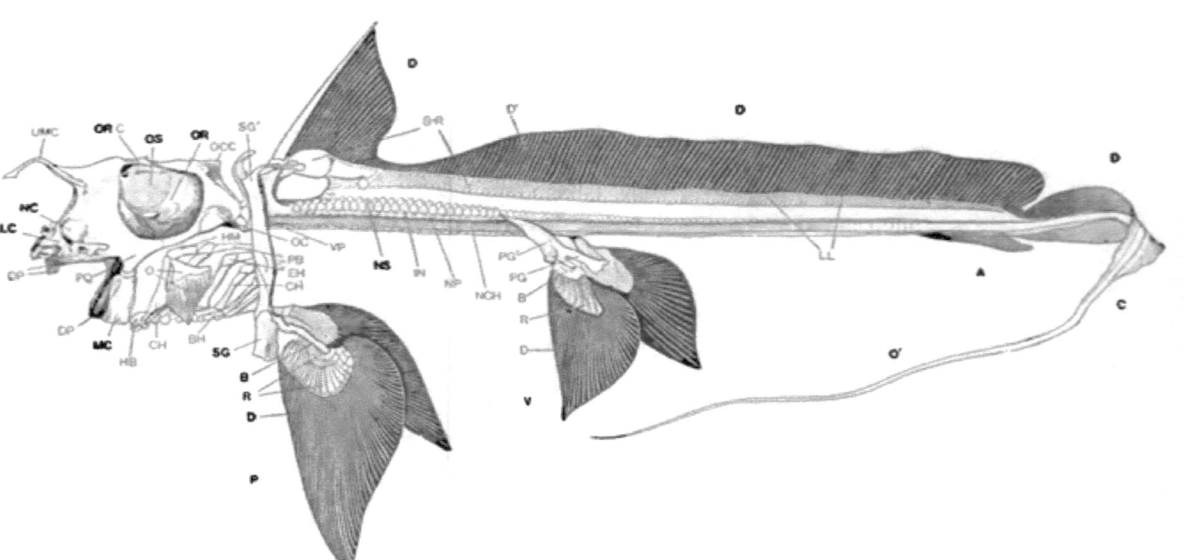

Fig. 105. — Skeleton of *Chimæra monstrosa*, ♀. (Drawn from a preparation of Fric by Arnold Graf).
A. Anal fin. B. Basal cartilaginous fin supports. B+R. Basal and radial fin supports. BH. Basihyal element of gill arch. C. Caudal fin. CH. Ceratohyal (and ceratobranchials). D. Dorsal fin, its foremost element a stout erectile spine. D'. Dermal fin rays. DP. Dental plates. EH. Epihyal. HB. Hypobranchial. HM. Hyomandibular. IN. Interneural. LC. Lablo-nasal cartilages. MC. Mandible (Meckel's cartilage). NC. Nasal capsule. NCH. Notochord, showing annular calcifications of sheath. NP. Neural processes. NS. Neural spine. O. Opercular cartilages. O'. Opisthure. OC. Occipital condyles. OCC. Occipital crest. OR. Orbit. ORC. Orbital crest. OS. Orbital septum. P. Pectoral fin. PB. Pharyngobranchial. PG. Pelvic girdle. PG'. Dorsal process of PG. PQ. Palatoquadrate. R. Radial cartilaginous fin supports. SG. Shoulder girdle. SG'. Dorsal process of SG. UMC. Upper median cartilage of snout (not the clasping spine of male, v. p. 101). V. Ventral fin. VP. Anterior vertebral plate.

102

morphosed ; the mandible appears to be *autostylic*, or articulated directly with the skull cartilage, *PQ*. The gill
arches are shark-like, but the hyoid arch appears far less
modified than in sharks ; its upper element, *HM*, is thus
unconnected with either the skull or the joint of the jaw ;
its distal element, *CH*, has, however, developed a series of
specialized supports for the dermal gill shield, *OP*. The
study of the fin supports shows the dorsal elements, $B+R$,
representing probably the radial and basal elements together, arranged in a single row margined distally by the
longitudinal ligament, *LL*, supporting the dermal functional fin, *D*. The paired fins are readily reduced to the
plan of those of Fig. 84 ; their girdles, however, seem to
have acquired more modified characters, their ventral and
dorsal elements greatly increasing in size.

Chimæroids as a group have received but a small share
of the attention paid to the other fishes ; their living
forms are few and comparatively rare ; their embryology
and larval history are unknown ; and their life habits have
been suggested only in the work of Dr. Günther (*Challenger Report*) His record of the taking of immature
specimens of *Chimæra* at great depths seems thus far the
most important clue as to the conditions of their living
and breeding.*

Fossil Chimæroids

Fossil Chimæroids have left behind them very imperfect
records of the history of their group. Like the sharks,
little more than their dental plates and fin spines have
usually been preserved. The structures of some of their
ancient members appear to have differed little from those
just described in the recent Chimæra. In *Ischyodus*,

* Cf. also Goode and Bean, on *Harriotta*, P. U. S. Nat. Mus., XVII. 471–473.

a Jurassic form (Fig. 105 *A*), the skeletal structures are readily comparable to those of Fig. 105. In the case of two of the Mesozoic genera, however, the evolution of the Chimæroids had evidently attained a high degree of specialization : *Myriacanthus* and *Squaloraja*, whose partial restoration has been attempted in Figs. 106 and 106 *A*, must be both looked upon as highly modified forms ; their snouts and frontal spines are greatly enlarged, and their dental plates (Figs. 107 and 108) widely divergent from the general Chimæroid type : in Myriacanthus a series of membrane bones occurs in the head region (Fig. 106, *B, C*). In Squaloraja a horizontally flattened body shape parallels the development of the ray-like form of sharks.

Fig. 105 A. — The Mesozoic Chimæroid *Ischyodus.*   × ¼.   (After ZITTEL.)

Living Chimæroids

The Chimæroids of to-day must be looked upon as the survivors of a group comparatively numerous in Mesozoic times : the few existing forms accordingly, from the palæontological standpoint, acquire an exceptional interest. They have been grouped under three genera, — *Harriotta, Callorhynchus*, and *Chimæra*. The first of these (Fig. 117, *A, B, C*) has been only recently discovered, and but a few examples have been taken ; it merits especial attention, since it is unquestionably the most shark-like of known Chimæroids. In the male it lacks entirely the frontal spine and has its claspers in an exceedingly un-

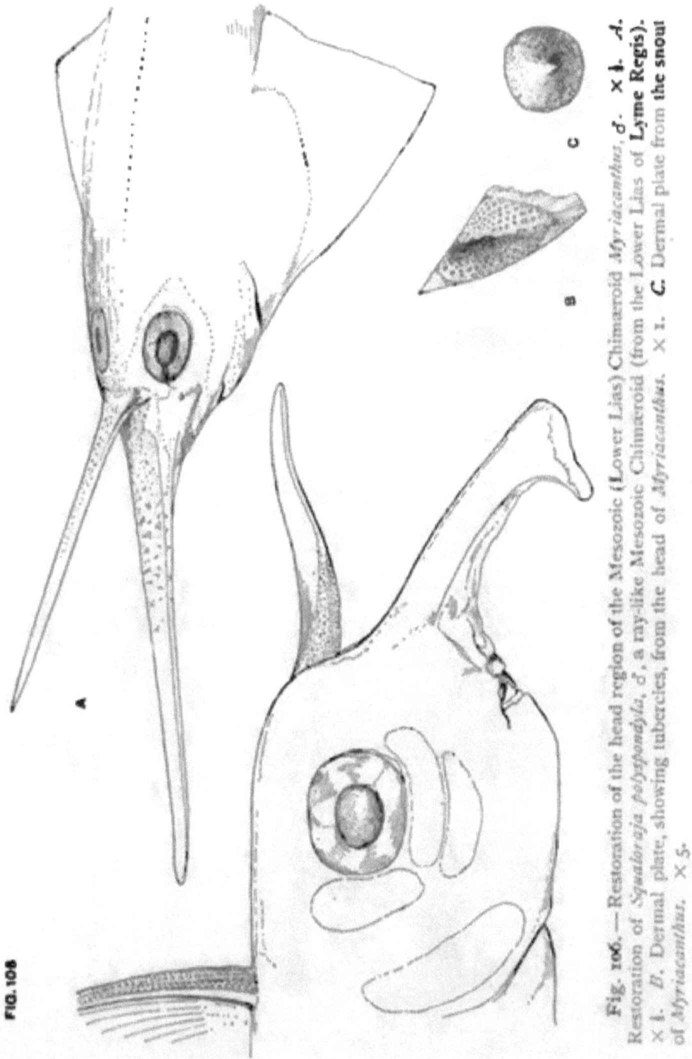

Fig. 106.—Restoration of the head region of the Mesozoic (Lower Lias) Chimæroid *Myriacanthus*, ♂. × ⅓. *A*. Restoration of *Squaloraja polyspondyla*, ♂, a ray-like Mesozoic Chimæroid (from the Lower Lias of **Lyme Regis**). × ⅓. *B*. Dermal plate, showing tubercles, from the head of *Myriacanthus*. × 1. *C*. Dermal plate from the snout of *Myriacanthus*. × 5.

differentiated **condition.** **The eggs** are evidently fertilized after they have been extruded.

The second genus, Callorhynchus, **is** represented **by but**

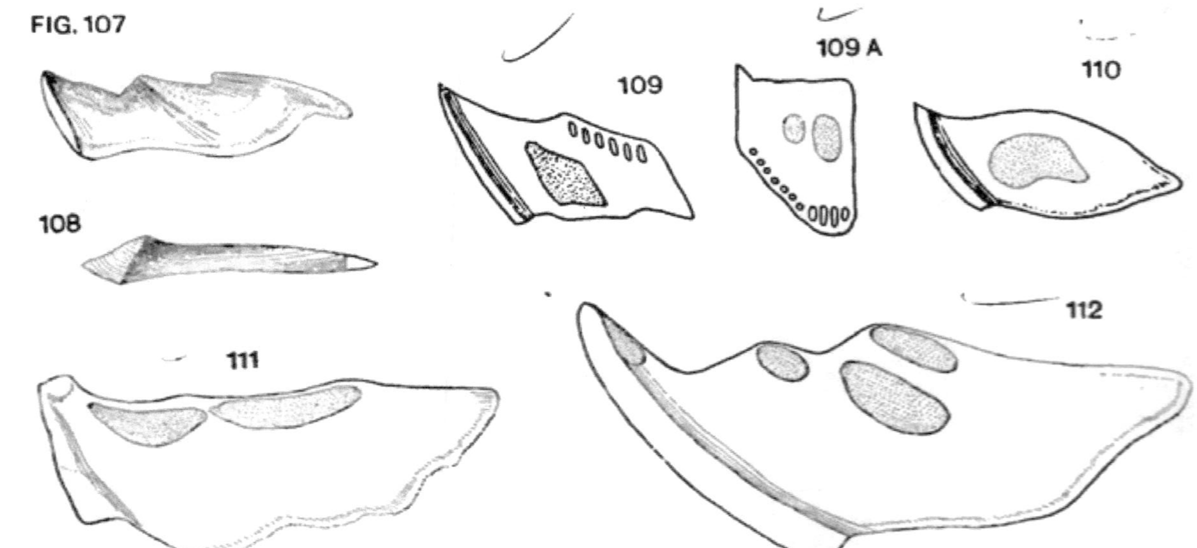

Figs. 107–112.—Dental plates of Chimæroids, showing inner aspect. The "tritors" are indicated by dotted areas. 107. *Myriacanthus*, right mandibular. (After Smith Woodward.) 108. *Squaloraja*, right mandibular. (After Smith Woodward.) 109. *Chimæra*, right mandibular. (From Smith Woodward, after Newton.) 109 A. *Chimæra*, left palatine plate, in ventral view. (From Smith Woodward, after Newton.) 110. *Callorhynchus*, right mandibular. (From Smith Woodward, after Newton.) 111. *Rhynchodus crassus*, right mandibular. (Middle Devonian.) (After Newberry.) 112. *Ischyodus Egerti*, right mandibular. (From Smith Woodward, after Newton.)

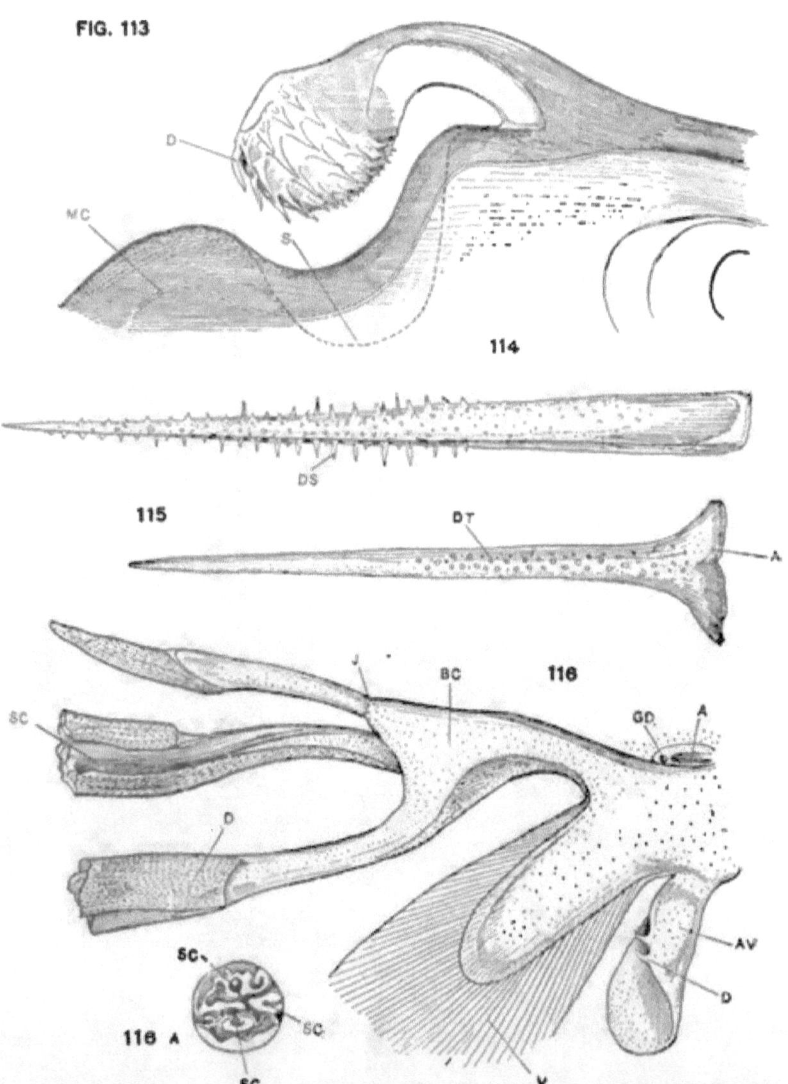

Figs. 113-116 A. — Spines and clasping organs of Chimæroids. 113. Clasping spine of the forehead of male *Chimæra colliei*. × 6. 114. *Myriacanthus* dorsal spine. (After L. AGASSIZ.) 115. Frontal spine of male *Squaloraja*. (From SMITH WOODWARD.) 116. Ventral fin and clasping organs of male *Chimæra colliei*. × 1. 116 A. View of tip of hinder clasper (intromittent organ), when the three tips are drawn together.

A. Anus. *AV.* Anterior rim of ventral fin, specialized as a clasping organ. *BC.* Body of the posterior clasper (intromittent organ). *D.* Dermal denticles. *DS.* Dermal spine-like denticles. *DT.* Dermal tubercles. *GD.* Urinogenital aperture. *J.* Jointed base of inner ventral element of intromittent organ. *MC.* Mucous canal. *S.* Sheath of frontal spine. *SC.* Sperm groove of inner face of clasper. *V.* Ventral fin.

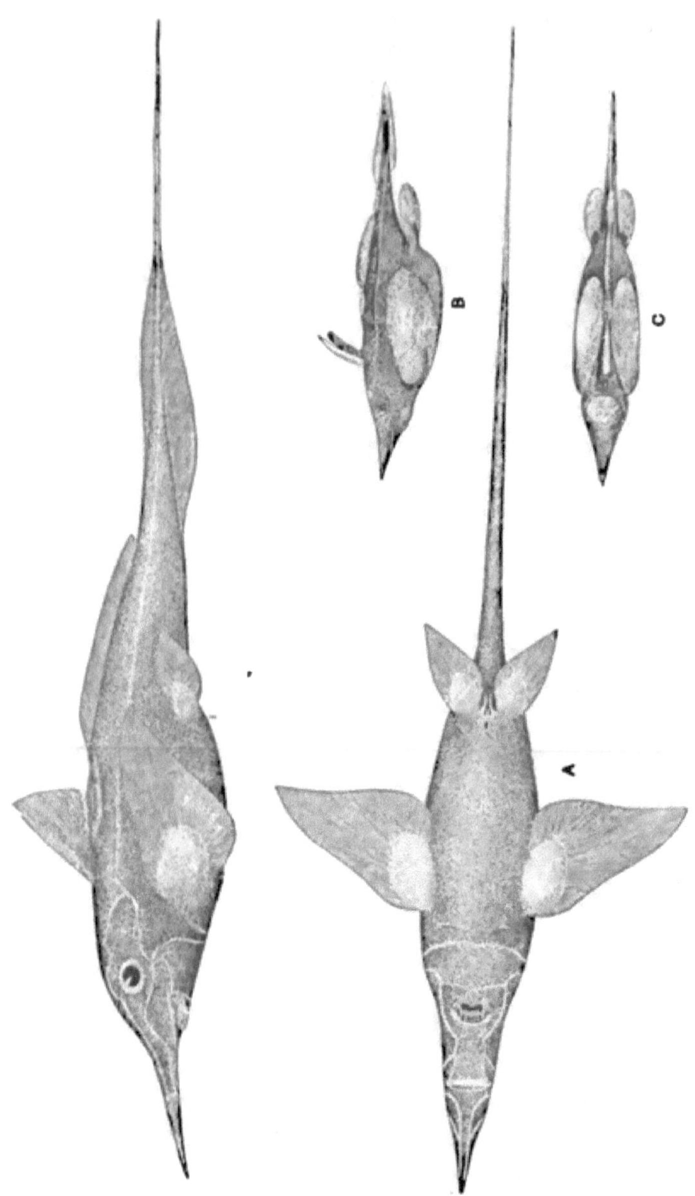

Fig. 117.—*Harriotta raleighana*, Goode and Bean. ♂. × ⅓. A new genus of Chimæroid — a bathybial form. *A.* Ventral view, showing rudimentary claspers. *B, C.* Immature specimens.

a single species, *C. antarcticus.* It is said to be common in the Straits of Magellan, and is popularly known as the Bottle-nosed Chimæra (Fig. 118, *A, B*). Its remarkable snout is well supplied with sense organs, and its pad-like dilation in front of the mouth is evidently of barbel-like function; it illustrates closely, no doubt, the remarkable snout process of Myriacanthus. Callorhynchus is shark-like in its general shape; and its caudal, dorsal and ventral

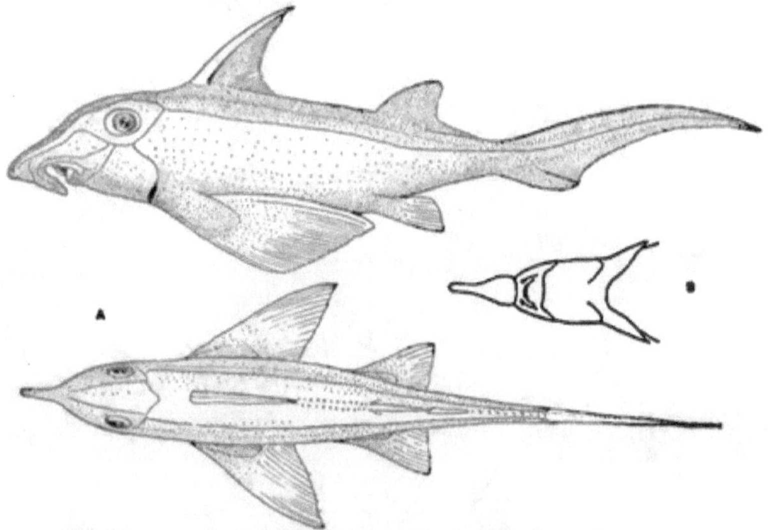

Fig. 118. — The bottle-nose Chimæra, *Callorhynchus antarcticus,* ♀. × ⅓. From Magellan Straits. *A.* Dorsal aspect. *B.* Ventral view of head. (After GARMAN.)

fins correspond closely in appearance and structure with those of certain sharks; the greatly enlarged pectoral fins have, however, a more highly specialized character; they stand boldly out from the sides of the body, and their bases are rounded and muscular. The mucous canals (Garman) have paralleled the saccular or tubular structures of the majority of sharks. The mandible (Fig. 110) shows but a single broad tritoral area.

Chimæra, the third genus of the recent forms, is well represented in the commoner form, *C. monstrosa* (Fig. 119, *A, B*). This species is widely distributed in the Mediterranean and Atlantic, taken usually in deep water; it is the largest of the living species, often attaining a yard in length. Its occurrence is usually erratic: in a favourable locality, as at Messina, months often elapse before one is taken; at other times many will be brought in in the course of a few days. The Portuguese species, *C. affinis*,

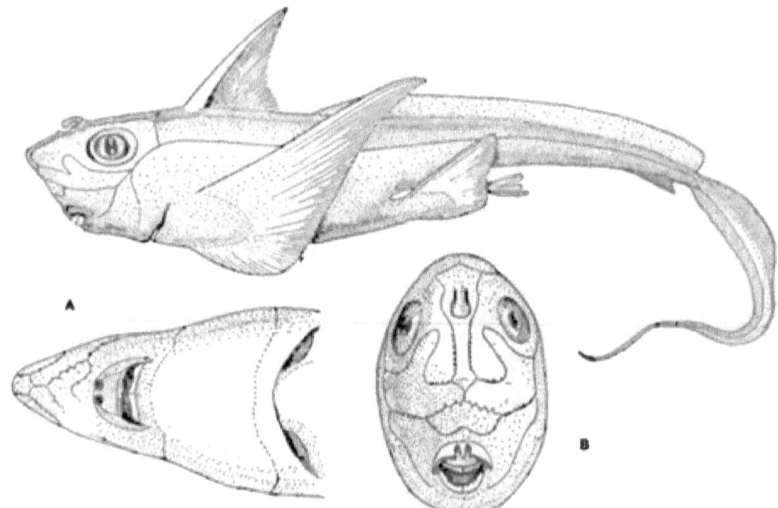

Fig. 119.—The sea-cat, *Chimæra monstrosa*, ♂. × ⅙. *A*. Ventral view of snout. *B*. Front view of head. (After GARMAN.)

is said to be numerous in the deep fishing grounds; the writer has seen it in the Lisbon market, where from its low price it evidently ranks with the sharks as a food fish. The smaller Pacific *C. colliei* (Fig. 104), rarely half a yard in length, differs sharply from the other species, and is therefore often given rank as a distinct genus, *Hydrolagus*, Gill. The writer learns from his friend Dr. Bean that it occurs abundantly in the shallow waters of Vancouver; it is there well known as the "rat fish," and may often be seen in the neighbourhood of the docks, swimming slowly at the surface.

The shape of the body of Chimæra seems in some regards to have diverged from the more shark-like form of Callorhynchus. Its organs have become concentrated in the pectoral region, and the disturbance in the curve normals of the fish seems to have caused the shortening of the snout, and the sudden dwindling of the hinder trunk region; the tail, with its thread-like terminal, the opisthure (Fig. 120), is accordingly to be looked upon as de-

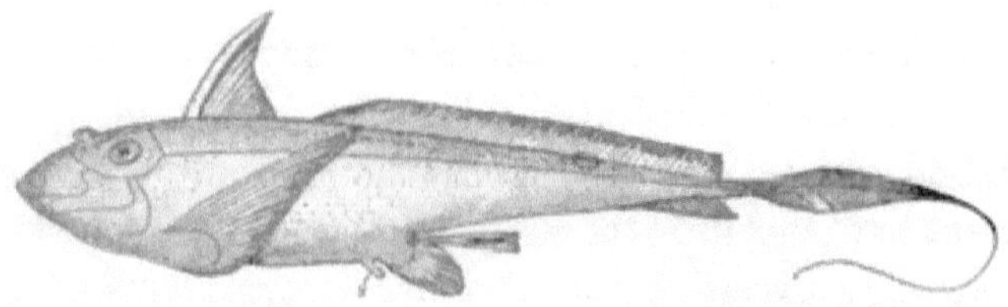

Fig. 120. — *Chimæra monstrosa*, ♂. Juv. × about ½. (After L. AGASSIZ.) The anterior ventral clasper is noted at *X*; the tail terminates in a thread-like opisthure.

generate. In the anterior region, however, a number of what seem to be primitive characters have been retained; the mucous canals are groove-like; and the dental plates (Figs. 109, 109 *A*) exhibit a series of tritoral areas.

Affinities

All that is known of Chimæroids, living or fossil, gives but little definite knowledge of the kinships or evolution of the group. Their shark-like structures cannot be shown to have taken their origin from shark-like conditions. Thus the dental plates even of the most ancient forms do not suggest their derivation from shagreen cusps; the beak-like jaws of the Devonian Rhynchodus (Fig. 111), of the Devonian Ptyctodus, or of the Mesozoic genera, *e.g.* Ischyodus (Fig. 112), differ little in their structures from those of their living kindred (Figs. 109, 109 *A*, 110). The tritors accordingly are only doubtfully to be derived from the fusion of the primitive basal substance of the teeth with the tissue of the jaws. But the history of Chimæroids tells of their ancient importance and of the diversity of their forms, and demonstrates that they cannot be connected with other existing forms of fishes. In Liassic times their specialized members bore the same relation to Chimæra as did the aberrant Cestracionts of the Coal Measures to the simpler sharks. In their dental evolution they had even reached a more specialized condition than the Cochliodonts (Cestracionts?). Thus in Myriacanthus and Squaloraja, "all anterior prehensile teeth have disappeared, and the growth of the dental plates, instead of taking place exclusively at the inner border, seems to have gradually extended to the whole of the attached surface. The Chimæridæ exhibit an advance in the circumstance that all the dental plates are thickened, while the hinder upper pair are both closely apposed in the median line and much extended backward" (Smith Woodward).* Squaloraja had certainly attained a high degree of evolution in the calcified vertebral rings, and in its specialized girdles, fins, and clasping organs. Myriacanthus, on the other hand, while retaining its ancient vertebral characters, had evolved a well-marked series of membrane bones.

One cannot deny that the study of Chimæroids as a group emphasizes many of their structural affinities to the sharks. They resemble them in their cartilaginous skeleton, fins and girdles, "claspers," integument, and

* *Cat. Fossil Fishes* II, xvi.

sense organs: they present similar **visceral** characters, spiral intestine, heart, gills, abdominal **pores, renal and** reproductive organs.

Their more important **divergences from the plan of** elasmobranchian structure may thus be summarized:—

I. Skull and mandible (v. pp. 252, 256). The mandible articulates directly with what appears to be the **carti**-lage of the cranium, *i.e.* without the hyoid-arch element serving as the suspensorium (*Autostylic*, p. 257).

II. Fins, paired (Wiedersheim) and unpaired (Ryder), and fin defences. The first dorsal, armed with an anterior spine, is so specialized that **it folds like** a fan, and may be depressed **into a receptive sheath. The tail is (second**-arily) diphycercal.

III. Skin defences **and teeth.** Shagreen **tubercles** occur in Chimæroids **and are** in every **way shark-like.** They are scattered thickly **over** the entire dorsal region in *Menaspis*,* sparsely in Squaloraja. They occur in the head region and on the spines in Myriacanthus (Figs. **106** *C*, **114**); **and** on the head, spine, and clasper tips of recent forms (Figs. 113 *D*, **116** *D*). **But** dermal bones also occur, as in Myriacanthus (Fig. 106 *B*), **which** do not outwardly resemble the structures of **ancient** sharks **shown,** *e.g.* in Fig. 90 *B*. **The** dermal plates protecting the suborbital sensory canal of Chimæra (Fig. 104, *DP*) must **be looked** upon as specialized defences, not as degenerate remnants of a complete dermal armouring (Pollard). And the dental plates, as already noted (p. 99), are altogether unshark-like; their tritors are few in number and constant in position, suggesting **an origin** from more superficial tooth centres, but these in turn, like the toothplates of Cestracionts, may have been evolved from shagreen **denticles.**

* Jaekel, *SB. d. Gesell. nat. Freunde,* Berlin, 1891, Nr. **7.**

I

IV. GILL ARCHES. The gills have become drawn closely together as in the more highly evolved types of fishes (*e.g.* bony fishes), and are enclosed by a protective dermal flap which fringes the sides of the head. The concentration of the arches and the appearance of the dermal shield suggest, however, the conditions we have seen in ancient sharks (Cladoselache, Chlamydoselache, Acanthodes), and cannot be given significance as the ancestral form of the opercular apparatus of Teleostome. Even the similar conditions of the Chimæroid and ancient shark may well have been evolved independently. It is interesting to note that in Chimæroids the spiracle is absent.

V. BRAIN. The brain structure is archaic. Its general plan is, however, more shark-like than Dipnoan (Wilder, *Ref.* p. 244).

VI. LATERAL LINE. The sensory canals possess many distinctive features; they retain their groove-like character, but become widely sacculated and dilated, especially in the snout region.

VII. CLASPING SPINE. The forehead clasper of the male has been a well-marked character of Chimæroids from Liassic time. It folds anteriorly into a receptive groove; its distal end, studded with recurved spines, serves in the recent forms for strongest retention. It seems to represent morphologically the anterior spine of a dorsal fin (cf. Pleuracanthus, p. 83).

In spite of these differences, however, the kinships of the Chimæroids seem unquestionably nearer the stem of the sharks than that of other fishes. On existing evidence the Chimæroid could not have been derived from either Teleostome or lung-fish; nor, on the other hand, could any of the larger groups of fishes be reasonably derived from its conditions as ancestral. The dentition of Chimæroids alone is so remarkable that no direct process of differentiation could convert it into the structures of lung-fish or Ganoid. A number of archaic features draw fishes together in the lines of their descent, but they cannot be interpreted as linking the Chimæroids with the Dipnoans, or the Dipnoans with the Chimæroids. Autostylism, often adduced to ally these groups, differs widely in its characters in each (p. 254) : and the apparent similarities in dental plates and membrane bones are closely paralleled by the sharks. The diphycercal tail of the Chimæroid can be made no standard of comparison, since it is evidently a secondary structure, arising within the limits of the group, as it may well have done among sharks (Pleuracanthus) or Teleostomes (Polypterus, eel).

If the sum of the general characters of Chimæroids be considered, their affinities would clearly be to the most ancient sharks. Their structures are not so widely at variance with those of Elasmobranchs that they cannot reasonably be derived from their more generalized conditions in vertebral characters, cranium, mandible, girdles, fins, membrane bones, gills. Absence of swim-bladder is again strikingly shark-like. Like the ancient sharks, they have been well adapted for survival by evolving but few specialized structures (*e.g.* dentition, gills). Their ventral clasping organs separate them clearly from the Dipnoans. Until the discovery of Harriotta the frontal clasping spine remained as one of the most distinctive features of Chimæroids ; its high degree of specialization in Liassic times is alone significant of the antiquity of their descent.

VI

THE LUNG-FISHES

Lung-fishes, or Dipnoans, have long been looked upon as the linking type between amphibians and fishes. In some regards of structure they approach the primitive sharks; in others, they resemble so closely the salamanders that they were recently regarded by W. N. Parker as worthy of a class by themselves, intermediate between fishes and amphibians. As with the Chimæroids, their few surviving members give but a mere suggestion of the former size and importance of the group.

Structural Characters

The general structural plan of a Dipnoan is shown in the adjoining figure (Fig. 121), taken from a dissection of the African form, *Protopterus*. Its thick, spindle-shaped body, enclosed in rounded, horn-like scales, *CS*, terminates in a diphycercal tail, *CF*. The head is salamander-like both in shape and in slimy integument. The paired fins (schematized in the figure, *PF, VF*) are archipterygial.

The head region is characterized by a cartilaginous brain case, roofed by dermal bones, *HR*; a mandible, *MA*, directly articulated with the skull (autostylic); an anterior and posterior nares, *NO*, — the former opening under the lip, the latter within the mouth; a row of small, compressed (unsegmented) gill arches, *GA*, whose single outer aperture

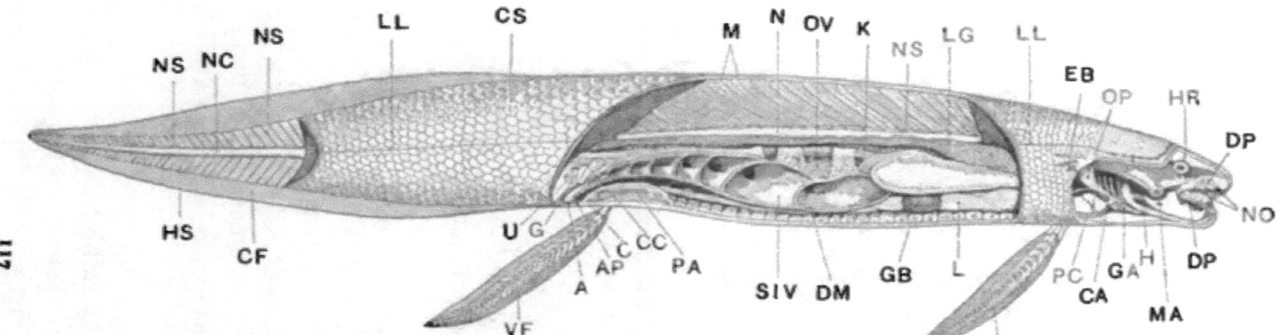

Fig. 121. —General anatomy of lung-fish.   (In the main after W. N. PARKER's figure of *Protopterus*).   The paired fins are schematized. *A.* Anus. *AP.* Abdominal pore. *C.* Cloaca. *CA.* Conus arteriosus. *CC.* Cloacal cæcum (rectal gland). *CF.* Caudal fin. *CS.* Cycloidal scales. *DM.* Dorsal mesentery. *DP.* Dental plates. *EB.* External gills. *G.* Genital duct. *GA.* Gill arches. *GB.* Gall bladder. *H.* Hyoid (ceratohyal). *HS.* Hæmal spines. *K.* Kidney. *L.* Liver. *LG.* Lung (air-bladder). *LL.* Lateral line. *M.* Muscle segments. *MA.* Mandibular (autostylic) articulation. *N, NC.* Notochord. *NO.* Anterior and posterior nares. *NS.* Neural spine. *OP.* Operculum. *OV.* Ovary. *PA.* Pelvic girdle. *PC.* Pericardium. *PF.* Pectoral fin. *SIV.* Spiral intestinal valve. *U.* Urinary duct. *VF.* Ventral fin.

117

is guarded by an operculum, *OP*. The stunted external gills which here protude, *EB*, are sometimes looked upon as significant of an ancestral condition (Garman, Wiedersheim).

The viscera are somewhat shark-like in their features. They include a short digestive tract, with well-marked spiral intestinal valve, *SIV*; a fenestrated dorsal mesentery, *DM*; a large, elongate liver, *L*; a heart whose arterial cone, *CA*, contains transverse rows of valves; a cloaca, abdominal pores (or pore), *A*; and a rectal cæcum, *CC* (v. p. 263). The elongate kidney, *K*, the ovary, *OV*, with its many small eggs, and the long, paired, sacculated air-bladder (lung) may be named as among the least shark-like of its visceral characters.

The skeleton of a Dipnoan (Fig. 122) is almost entirely cartilaginous. A stout notochord, encased in a heavy sheath, *NCH*, passes from the skull to the tip of the tail: vertebral centra encroach upon it only in the caudal region, *C*. Dorsal and ventral processes, arranged in metameral sequence, extend from the notochordal sheath outward and become distally the cartilaginous supports of the dermal unpaired fin. The proximal elements might thus be regarded as neural, *N*, *NS*, or hæmal processes and spines, the distal elements as equivalent to the basal and radial fin supports, $B + R$. A stout, longitudinal ligament, *LL*, serves to connect the outer ends of the cartilaginous processes, as well as the proximal ends of the dermal fin rays. The ribs are probably the homologues of the hæmal processes; the most anterior pair, greatly enlarged, extends downward on either side as the occipital ribs, *OR*, specialized in the function of the air-bladder.

The structure of the paired fin is normally of the archipterygial form of Fig. 54. In Protopterus, however,

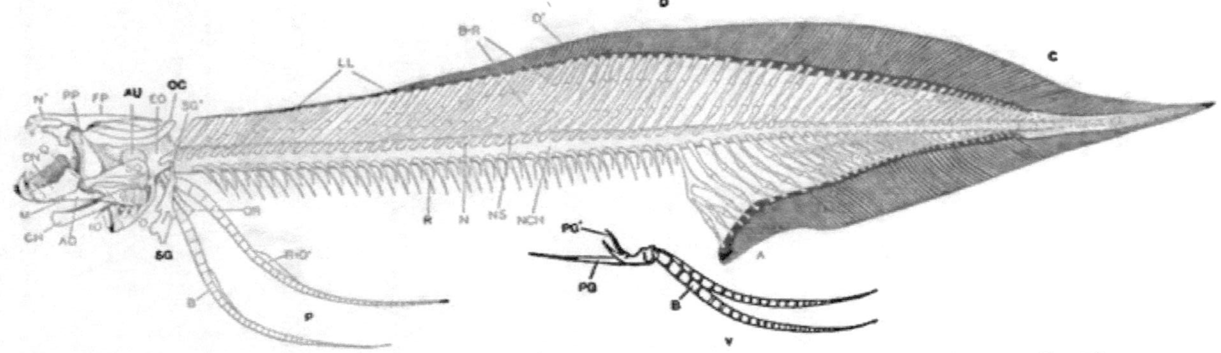

Fig. 122.—Skeleton of lung-fish, *Protopterus annectans*.   (From preparation of FRIC; drawn by Dr. ARNOLD GRAF.)
   A. Anal fin region. *AG.* Angular. *AU.* Auditory capsule. *B.* Basal cartilaginous fin supports. *B+R.* Basal and radial supports. *C.* Caudal. *CH.* Ceratohyal. *D.* Dorsal fin region. *D'.* Dermal fin rays. *DN.* Dentary. *EO.* Epiotic. *FP.* Frontoparietal. *IO.* Interoperculum. *LL.* Longitudinal ligament. *M.* Meckel's cartilage. *N.* Neural process. *N'.* Nasal. *NS.* Neural spine. *O.* Operculum. *OC.* Occipital crest. *OR.* Occipital rib. *P.* Pectoral fin. *PG.* Pelvic girdle. *PG'.* Dorsal process of *PG.* *PP.* Palatopterygoid. *Q.* Quadrate. *R.* Radial. *R+D'.* Radial and dermal fin elements. *SG.* Shoulder girdle. *SG'.* Dorsal process of shoulder girdle. *SQ.* Squamosal. *V.* Ventral fin.

(Fig. 122), this plan of structure is somewhat obscured by the rudimentary character of the radial and basal elements, $R + D$, although the fin stem presents a well-marked jointed character, B. The pelvic girdle, a solid plate of cartilage, is produced anteriorly into a narrow median outgrowth, PG, and laterally into a pair of dorsal spurs, PG'. The shoulder girdle is composed on either side of a large ventral element, SG, which meets its fellow in the median ventral line, and of a short dorsal element, SG', which connects it with the skull.

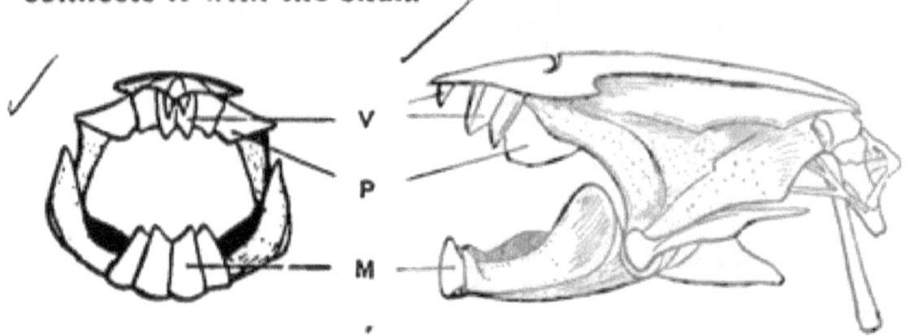

Fig. 122 A. — Jaws and skull of *Protopterus annectans*, figured in front and side aspects, showing paired dental plates. ×⅓. (After NEWBERRY.)
M. Dental plates of (dentary) mandible; *P.* of palatopterygoid; *V.* of vomer.

In the head region (v. pp. 252, 254), the brain case is cartilaginous, with, however, a few true bone centres (*e.g.* epiotic) appearing; the roofs of the skull and mouth, together with the mandible, are well sheathed by dermal bones, as FP, N, PP, DN, AG. Paired dental plates fringe the rim of the mandible (Fig. 122 A, M), the vomerine region (V), and the anterior end of the palatopterygoids (P).

Fossil Lung-fishes

The structures of the recent Dipnoans can as yet be but imperfectly compared with those of fossil forms. Their ancestral conditions can only be determined when more

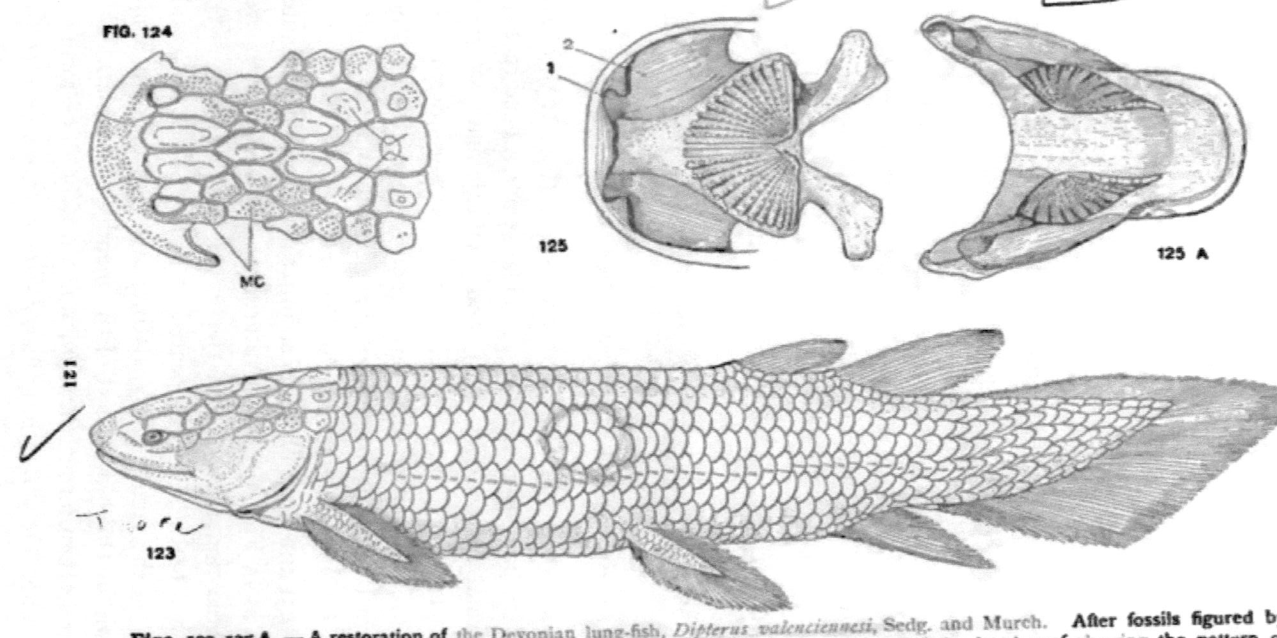

Figs. 123-125 A. — A restoration of the Devonian lung-fish, *Dipterus valenciennesi*, Sedg. and Murch. **After fossils figured by** Smith Woodward and by Pander. × about ⅓. 124. *Dipterus;* the dermal bones of the head roof, showing the pattern of arrangement characteristic of lung-fishes. *MC.* Openings of mucous tubules. (After Smith Woodward.) 125. Upper, and 125 *A.* lower jaw of *Dipterus.* 1 and 2 indicate the position of the anterior and posterior nares. (After Smith Woodward.)

perfect evidence is discovered as to their kinship and the lines of their descent.

In the history of fishes, Dipnoans are known to have been early a dominant group. In some regards, one of their ancient forms bore many resemblances to the Pleuracanthid shark, which, although known at present only in a later period, may well have been its contemporary. But the range in the forms of Dipnoans occurring in the early Palæozoic indicates the remote antiquity of their origin. They had even then evolved exoskeletal characters which are scarcely less specialized than those of existing forms.

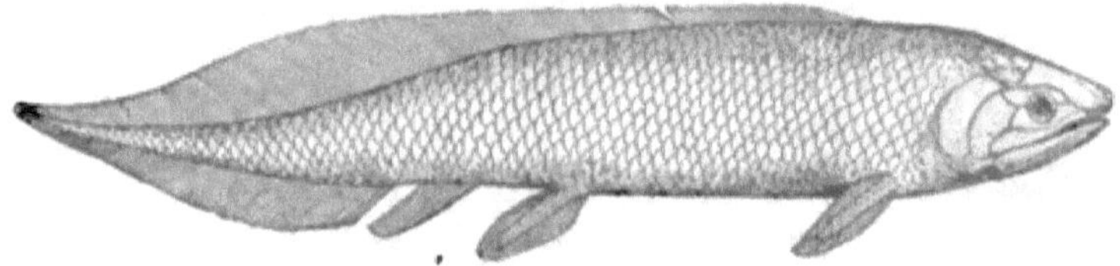

Fig. 126. — A restoration of the Devonian lung-fish, *Phaneropleuron.* × ⅓.

Dipterus, of the Old Red Sandstone (Fig. 123), had a complete body armouring of cycloidal scales, a head roofing of dermal plates (Fig. 124), and well-calcified jaw rims (Figs. 124, 125, 125 A). Its fin rays were dermal in structure, its paired fins were archipterygial, its tail and its dorsal fins separate and lobate. Its mucous canals had become elaborately adapted to the body scales (lateral line, Fig. 123) and head plates, piercing the latter with minute pores, as in Figs. 65, 66. Anterior and posterior nares are indicated under the rim of the upper jaw (Fig. 125, 1–2). Marginal teeth have disappeared; a pair of elaborate dental plates on the mouth roof (palatine) are apposed by a similar pair in the hinder part of the mandible (splenial).

The Carboniferous *Ctenodus* was a nearly allied form.

Another Devonian lung-fish, *Phaneropleuron* (Fig. 126),

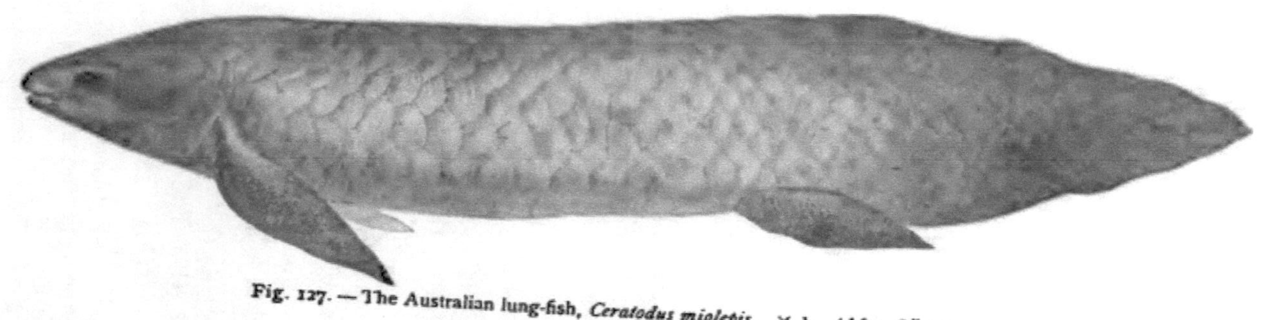

Fig. 127. — The Australian lung-fish, *Ceratodus miolepis.* × ⅓.   (After GÜNTHER.)

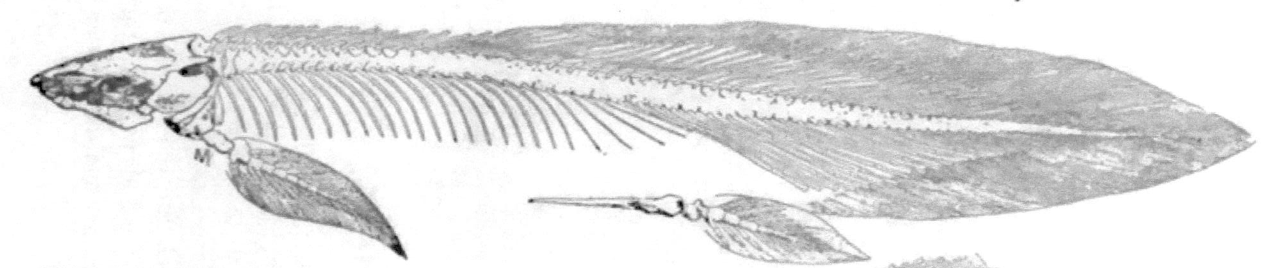

Fig. 128. — Skeleton of *Ceratodus.*   (After GÜNTHER.)   The vertebral structure of the tail region is shown in detail immediately below the caudal fin.

was similar to Dipterus in its skeletal characters. Its elongate diphycercal tail was continuous with the dorsal and anal (?) elements; in this, and in the retention of marginal cusp-like teeth, it resembled the Pleuracanthid sharks.

Living Forms

The three forms of living lung-fishes may reasonably be looked upon as the survivors of the more generalized Palæozoic forms. *Ceratodus*, the Australian genus, appears to have retained most perfectly the ancestral conditions; it has probably remained almost unmodified from the early Mesozoic times,[*] and presents close affinities to the Coal Measure family, *Ctenodontidæ*, and even to the Devonian Dipterids. Its outward appearance is shown in Fig. 127, and its skeleton in Fig. 128. The latter is seen to resemble closely the characters of Fig. 122; its paired fins are archipterygial; the mouth is lacking in marginal cutting plates (cf. *V*, Fig. 122 *A*). The dental plates

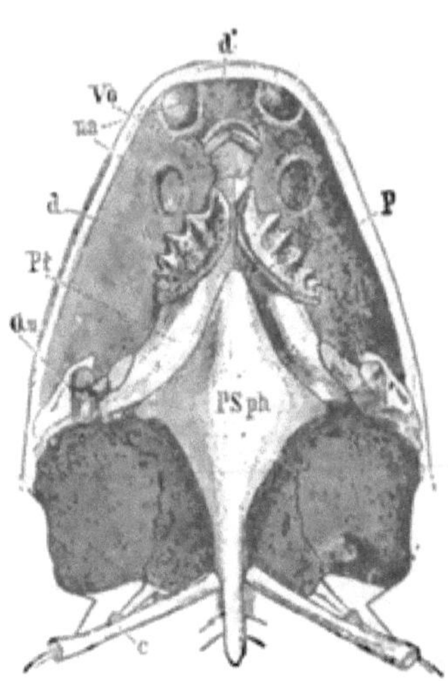

Fig. 128 A. — Skull of *Ceratodus*. Seen from the ventral side. (After ZITTEL.)

c. Occipital rib. *d.* Dental plates. *na.* Anterior and posterior nares. *P.* Palatine. *PSph.* Paraspenoid. *Pt.* Pterygoid. *Qu.* Quadrate. *Vo.* Vomer.

of the palatine and splenial regions (Fig. 128 *A*) are seen to correspond clearly with those of Figs. 125, 125 *A*.

Ceratodus had long been known to the colonists of

---

[*] v. p. 10.　The recent genus, according to Dr. Gill, is to be distinguished as *Neoceratodus*.

Queensland as a plentiful food-fish, a "salmon" in size and taste, although, curiously enough, it remained undescribed until 1870 (Krefft, and Günther). After this its developmental history was eagerly awaited, in the hope that it would reveal the affinities of the Dipnoans to the sharks, amphibians, and in general to the early chordates. About ten years ago Caldwell was sent to Queensland by the

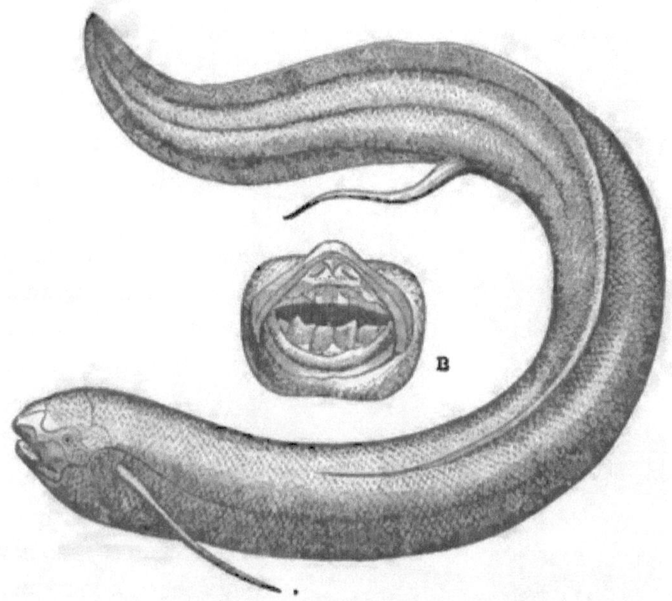

Fig. 129. — The South American lung-fish, *Lepidosiren paradoxa*, Natter. × ⅓. (From NICHOLSON, after NATTERER.) A front view of the mouth is shown at *B*.

Royal Society, and succeeded in securing a set of the embryonic stages, but his results still remain unpublished. A second set of embryos was collected in 1891 by Semon, from whose recent paper a summary is later given (p. 198). The development of Ceratodus, however, as far as it is at present known, has proven in many ways unsatisfactory to the phylogenist; its abbreviated growth stages cannot be

looked upon as furnishing clearly the ancestral history of
Dipnoans.

The two remaining forms of recent lung-fishes, *Lepidosiren* and *Protopterus*, resemble each other so closely that Ayers has contended that they should be regarded as distinct only specifically.

Lepidosiren, the South American form (Fig. 129), was discovered by its describer, Natterer, in 1837 in the upper Amazon. It then, for many years, succeeded in eluding the collectors, and was known as one of the rarest specimens of foreign museums. In 1887 it was, however, rediscovered in Paraguay, where it appears to have long been known as a food-fish. Its structures are now regarded as entitling it unquestionably to the rank of a distinct genus.

Protopterus, common in the White Nile and Congo (Fig. 129 *A*), has long been the "Lepido-

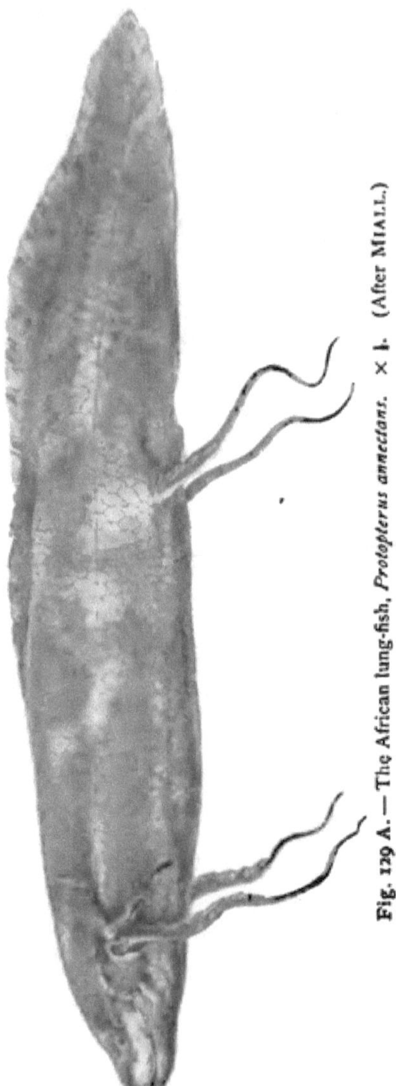

Fig. 129 A.—The African lung-fish, *Protopterus annectans*. × ½. (After MIALL.)

siren" of dealers, often of museums. It is the best known of Dipnoans, on account, partly, of the ease with which it may be transported alive. In the hardened mud cocoons with which it encases itself during the dry season, it is readily dug out of the stream bed and packed for exportation. When placed in tepid water, the cocoon dissolves and the fish shortly revives.

Relationships

A review of our knowledge of Dipnoans gives but little satisfactory suggestion as to their relations as a group. They must certainly be looked upon as an advancing phylum from which the amphibia may early have diverged. Their many amphibian characters have been lately emphasized by W. N. Parker. On the other hand, the evidences of the kinship of Dipnoans to the other types of fishes can only be interpreted as the common convergence of the ancient phyla toward the structures of the ancestral form of fish. Thus we find that the types of Devonian lung-fishes can only be distinguished from those of the contemporary Teleostomes by the pattern of arrangement of the plates of the head roof,* a condition which has led Smith Woodward to believe that these groups had already diverged before the appearance of dermal bones.

Lung-fishes have unquestionably many structures which may have been derived from the more generalized conditions of the sharks ; and as a group they may not unreasonably be looked upon as descended from the primitive elasmobranchian stem. Their ties of kinship to the sharks

---

* The present writer regards this distinction as somewhat provisional; median head plates are nominally characteristic of Dipnoans (Fig. 124), but, as in the sturgeons and siluroids, they are also well known among Teleostomes. Protopterus has, moreover, a symmetrical arrangement of the head plates.

have now been closened by the proof that their paddle-shaped fins may be directly deduced from a "monoserial archipterygium," and that their diphycercal caudal, formerly regarded as most primitive in plan, may have been acquired secondarily after a condition of heterocercy (W. N. Parker, Traquair, Dean).

The resemblances of Dipnoans to Elasmobranchs might be summarized in the following structures : — .

I. VERTEBRAL AXIS. Its notochordal condition and simple metameral, neural, and hæmal elements suggest the conditions of Cladoselache (p. 80); in that ancient form, however, the vertebral processes had not come into relation with the unpaired fins.

II. SKULL. The chondrocranium is as yet largely retained; as yet no dentigerous membrane bones of the mouth rim (maxillary and premaxillary) have appeared.

III. TEETH. These are clearly of an elasmobranchian order; the tubercles of the dental plates (Fig. 125) suggest closely a shagreen pattern; in Phaneropleuron, marginal cusps have even been retained. The palatine and splenial plates parallel strikingly some of the forms of Cestraciont dentition.

IV. BRAIN. Its structures are of an advancing elasmo-branchian order, annectent with reptilian (Ceratodus) and amphibian types (Protopterus).[*]

V. VISCERAL CHARACTERS. Heart, gills, digestive tract, vessels, mesenteries.

The closely corresponding characters of Phaneropleuron and Pleuracanthus might be looked upon as independently acquired; but in view of the many nearnesses of their phyla, these characters may reasonably be regarded as proof of genetic kinship.

* Burckhardt.

The advancing structures of the Dipnoan include, in addition : —

I. EXOSKELETAL SPECIALIZATIONS. Head-roofing dermal bones (cf., however, Pleuracanthid) and cycloidal scales. In early forms (Dipterus) these appeared at the surface and were apparently enamelled. In recent forms they are deeply sunken in the integument (Prototerus). They suggest closely the structures of Crossopterygian (p. 149).

II. ARTICULATION OF THE MANDIBLE. This is autostylic, somewhat as in Chimæroid (v. p. 256). Its homology is obscure.

III. AIR-BLADDER. (v. p. 264).

IV ABSENCE OF VENTRAL "CLASPERS" (cf., however, Cladoselache).

V. TRUE POSTERIOR NARES (amphibian).

VI. THE GREAT SIZE OF THE CELLULAR ELEMENTS OF ALL TISSUES (amphibian); THE GLANDULAR STRUCTURES OF THE EPIDERMIS (amphibian).

VII. CIRCULATORY CHARACTERS: the three-chambered heart ; aortic arches.

VIII. LIMB STRUCTURE. This, however, is not to be interpreted as in any way directly transitional to cheiropterygium.

The Arthrodiran Lung-fishes

The ARTHRODIRA, as Smith Woodward has shown, may provisionally be regarded as an order of extinct and highly specialized lung-fishes. They occur geologically among the earliest fishes, and include a number of (Devonian) forms whose peculiar characters and gigantic size must have made them among the most striking members of ancient fauna. The group might be regarded as standing in the same relation to the ancient Dipnoans as Acanthodians to the Cla-

K

dosclachian sharks.   As recently as 1887 its members were associated by Traquair with Pterichthys, but the discovery of jaws, specialized dentition, fin spines, and highly evolved pelvic fins at once separate this group from the lowly Ostracoderms.

American Arthrodirans, described mainly by Newberry and by Claypole, have proven of especial interest.   They occur from the Silurian to the Coal Measures.   The giant predatory member of this group, *Dinichthys* (Frontispiece, and Figs. 133–137), attained a length of ten feet.   *Titanichthys*, less formidable in armour and dentition, may well have been twenty-five feet in length.   These forms occur almost exclusively in the Waverly of Ohio.   Their discovery has here been due to the efforts of Dr. William Clark of Berea, Rev. William Kepler of New London, and Mr. Jay Terrell of Linton ; and most of the type specimens have been preserved in the museum of Columbia College, New York.

The European member of this group is a small, fresh-water (?) form, *Coccosteus*, especially abundant in the Old Red Sandstone of Scotland.   It has thus far yielded the most complete material for study, and its structural characters might accordingly be described, since they are probably common to all members of the group.

The lateral view of Coccosteus is shown in Fig. 130, the dorsal aspect of the anterior region in Fig. 131, and the ventral view of the visceral region in Fig. 132.   It will accordingly be seen that the general shape of the body of this Arthrodiran was somewhat depressed ; that the head, shoulder, and stomach regions were protected by bony plates ; and that the trunk region was lacking in armouring, and short in relative length.   In well-preserved fossils the space occupied by the notochord, *N*, is seen to pass from the region of hinder plates of the body armour to that of the tip of the tail. This is seen to be bordered by neural and hæmal processes, *N, H*, which in size and character are somewhat comparable with those of Protopterus or Pleuracanthus. The dorsal fin presents a metameral series of supporting cartilages (radial and basal, *DR, DB*). The basal supports of each pelvic fin have become compressed into a flattened plate, *VB*. Pelvic fins were present, but there have as yet been found no traces of pectoral appendages. In Dinichthys Newberry believed that a pectoral fin spine was present, and that this fin was

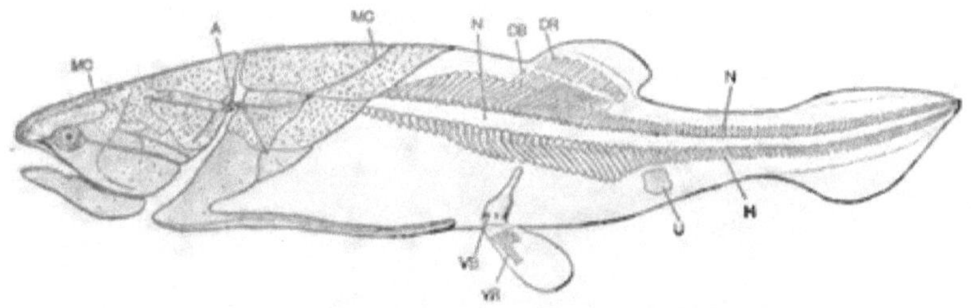

Fig. 130. — The Devonian Arthrodiran, *Coccosteus decipiens*, Ag. × ⅓. Old Red Sandstone, Scotland. (Side view, restored; slightly modified, after SMITH WOODWARD.)

A. Articulation of head with trunk. *DB.* Cartilaginous basals of dorsal fin. *DR.* Cartilaginous radials of dorsal fin. *H.* Hæmal arch and spine. *MC.* Mucous canals. *N.* Neural arch and spine. *U.* Median unpaired plate of hinder ventral region. *VB.* Basals of ventral fin. *VR.* Radials of ventral fin.

probably Siluroid-like (p. 171), but this view has not been confirmed.

The head of Coccosteus was clearly flattened, with orbits and nasal openings near its anterior margin; it was roofed by a stout buckler of closely fitted dermal plates (Fig. 131), whose outer surface was tuberculate, enamelled, and furrowed by sensory grooves, *MC*. The arrangement of the dermal plates of Coccosteus was early (1861) compared by Huxley with that of recent Siluroids,

an analogy afterward supported by Newberry, Dean, and
recently, on account of the similar characters of the sensory canals, by Pollard. In their conclusions, however,
fundamental characters of structure seem to have been
overlooked in the unlikeness of Arthrodiran to Teleostome. The inner structure of the cranium of Arthro-

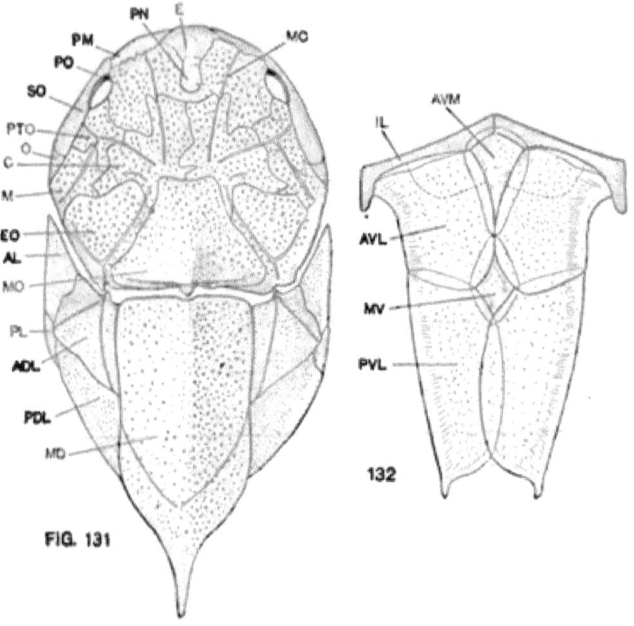

Figs. 131, 132. — *Coccosteus decipiens.* Dorsal view of dermal armouring. × ⅓.
(After TRAQUAIR.) 132. Ventral plates. (After TRAQUAIR.)
ADL. Antero-dorso-lateral. *AL.* Anterolateral. *AVM.* Antero-ventro-lateral.
C. Central. *E.* Ethmoid. *EO.* Epiotic. *IL.* Inferior lateral. *M.* Marginal.
MC. Mucous canals. *MD.* Median dorsal. *MO.* Median occipital. *MV.* Median ventral. *O.* Opercular. *PDL.* Postero-dorso-lateral. *PL.* Posterior lateral.
PM. Premaxillary. *PN.* Pineal. *PO.* Preorbital. *PTO.* Postorbital. *PVL.*
Postero-ventro-lateral. *SO.* Suborbital.

dira was evidently entirely cartilaginous ; in a Russian
Coccosteid, according to Smith Woodward, the base of the
brain case (parachordal cartilages) has been preserved and
shows a " tubular canal originally occupied by the anterior

extremity of the notochord." Gill arches and opercula are not definitely known. The mandible was attached directly to the skull (autostylic). The jaws were shear-like, their margins usually with pointed teeth, whose bases fuse with the tissue of the jaw and constitute dental plates. In all forms, as in Dinichthys (Frontispiece), there appear to have been three pairs of these "plates," those forming the rim of the mandible below, and those of the vomerine and palatine regions ("premaxillary" and "maxillary") above.* This arrangement of the dental plates somewhat resembles the Dipnoan's. Those of the Arthrodiran, however, appear to have been movable, and suggest a dental condition elsewhere unknown among vertebrates.

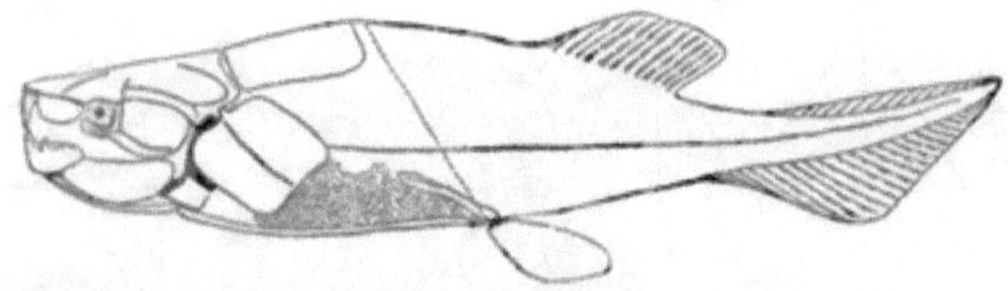

Fig. 133.—Restoration of *Dinichthys intermedius*, Newb. × ⅙. Cleveland Shales, Ohio.

The body armouring of dermal plates is characteristic of the group. A carapace, cape-like in shape, begins at the head angle and broadens out backward and dorsally towards the median line. It consists of a single median spade-shaped element, which forms the strong ridge of the back, and a flanking of lateral plates, all compactly joined. The rigid shield that is thus formed is movably connected with the head ; an elaborate joint, formed on either side between the anterolateral dorsal plate, Fig. 131, *ADL*, and the "epiotic," *EO*,—whence the name Arthrodira,—must

* According to Dr. Clark, an additional symphysial pair of dental plates was present in both upper and lower jaw (Dinichthys).

have permitted the head to be thrown backward to a
degree which suggests the thoracic joint of an Elater.
On the ventral side of the trunk there occurs a flattened
plastron (Fig. 132): its dermal elements are connected by
overlapping margins; they are lighter, and in some forms
(Fig. 135) lack the tuberculate surface of the dorsal
plates. Dorsal and ventral shields are connected by stout
lateral elements (Fig. 132, *IL*), which, passing ventrally,

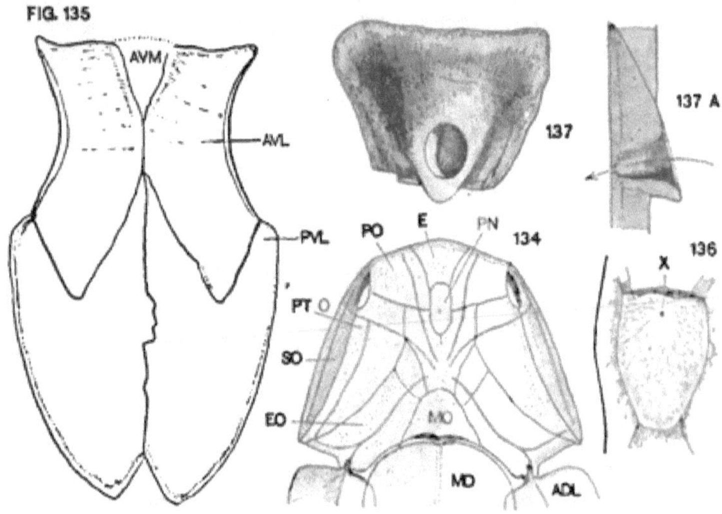

Figs. 134-137. — Dermal plates of *Dinichthys*. 134. Associated plates of head
and shoulders. 135. Plates of ventral armouring. (After A. A. WRIGHT). 136.
Pineal plate of *Dinichthys intermedius*, surface view. 137. Pineal plate of *Dinichthys terrelli*, visceral aspect. 137 A. Pineal plate, in sagittal section.

ADL. Antero-dorso-lateral. *AVL.* Antero-ventro-lateral. *AVM.* Antero-
ventro-median. *E.* Ethmoid. *EO.* Epiotic. *MO.* Median occipital. *PN.*
Pineal. *PO.* Preorbital. *PTO.* Postorbital. *PVL.* Postero-ventro-lateral. *SO.*
Suborbital. *X.* External aperture, and →, the axis of the pineal funnel.

meet in the median line, and become the anterior supporting rim of the plastron. By some writers these have been
homologized as "clavicles."

In further detail little is known of the anatomy of

Arthrodirans. Sensory canals have been described channelling the surface of the dermal plates of the dorsal side. In the body region of Coccosteus evidence of a lateral line occurs (Smith Woodward) in a white calcified band fossilized in the region of the space of the notochord. In this form, too, an endoskeletal plate is known, (Fig. 130, *U*) occurring in the median line in the region of the vent, which must be regarded as "suggesting an internal element of support occurring in the vertical septum between the right and left halves of some paired organ (*S. W.*)." The character of the dermal investiture of the trunk has apparently not been described; it may therefore be of interest to note that the museum of Columbia College has recently acquired two of the hinder dorsal plates of Dinichthys which clearly indicate the presence of integument. The plates are covered by a crinkled epidermis, whose irregular surface traceries resemble the roughened finish of Turkey morocco. This leather-like surface is seen to have been continued over the margin of the plates along the side of the trunk; traces of scales or tubercles are altogether lacking, and its appearance suggests that it may have been degenerate in structure.

Among Arthrodirans there occurs a series of most interestingly evolved forms; and it is found more and more evident that they, with other lung-fishes, may have represented the dominant group in the Devonian period, as were the sharks in the Carboniferous, or as are the Teleosts in modern times. There were forms which, like Coccosteus, had eyes at the notches of the head buckler; others, as *Macropetalichthys*, in which orbits were well centralized; some, like Dinichthys and Titanichthys, with the pineal foramen present; some with pectoral spines (?); some with elaborately sculptured derm plates. Among their forms appear to have been those whose shape was apparently sub-cylindrical, adapted for swift swimming; others (*Mylostoma*) whose trunk was depressed to almost ray-like proportions. In size they varied between that of a perch and that of a basking shark. In dentition (Figs. 138–144) they presented the widest range in variation, from the formidable shear-like jaws of Dinichthys to the lip-like mandibles of Titanichthys, the tearing teeth of *Trachosteus*, the wonderfully forked, tooth-bearing jaw tips of *Diplognathus*, to the Cestraciont type, Mylostoma. The latter form has hitherto been known only from its dentition, but now proves to be, as Newberry and Smith Woodward suggested, a typical Arthrodiran.

The puzzling characters of the Arthrodirans * do not seem to be lessened with a more definite knowledge of their different forms. · The tendency, as already noted, seems to be at present to regard the group provisionally as a widely modified offshoot of the primitive Dipnoans, basing this view upon their general structural characters, dermal plates, dentition, autostylism. But only in the latter regard could they have differed more widely from the primitive Elasmobranch or Teleostome, if it be admitted that in the matter of dermal structures they may clearly be separated from the Chimæroid. It certainly is difficult to believe that the articulation of the head of Arthrodirans could have been evolved *after* dermal bones had come to be formed, or that a Dipnoan could become so metamorphosed as to lose·not only its body armouring

* The writer believes that the Arthrodirans may as well be referred to the sharks as to the lung-fishes; as far as existing evidence goes, they certainly differed more widely from the lung-fishes than did the lung-fishes from the ancient sharks. They may, perhaps, be ultimately regarded as worthy of rank as a class.

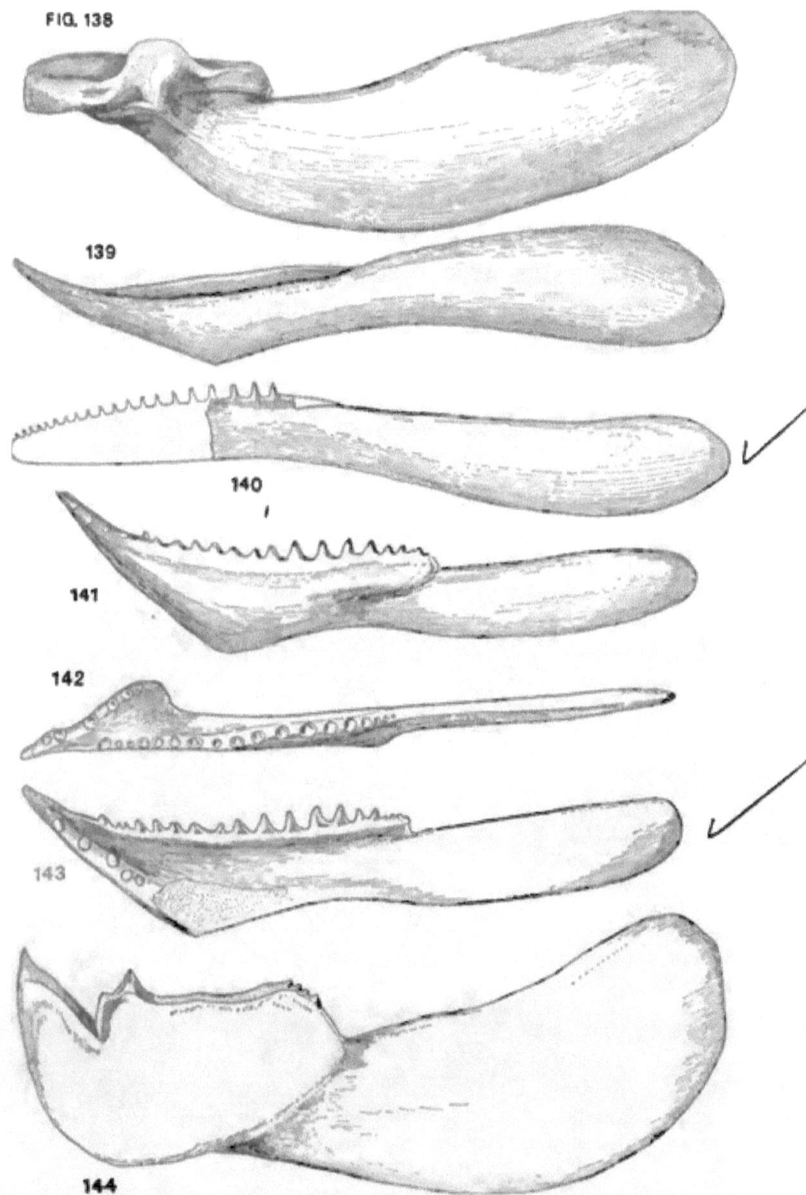

Figs. 138-144. — Mandibles of Arthrodirans: Cleveland Shale, Ohio. 138. *Mylostoma variabilis*, Newb., visceral aspect. 139. *Titanichthys clarki*, Newb., visceral aspect. × ¼. 140. *Trachosteus*, Newb., outer aspect. × ¼. 141. *Diplognathus*, Newb., outer aspect. × ¼. 142. *Diplognathus*, seen from dorsal side. 143. *Diplognathus*, visceral aspect. 144. *Dinichthys intermedius*, outer aspect. × ¼.

but its pectoral appendages as well. The size of the pectoral girdle is, of course, little proof that an anterior pair of fins must have existed, since this may well have been evolved in relation to the muscular supports of plastron, carapace, trunk, and head. The inter-movement of the dental plates, seen especially in Dinichthys, is a further difficulty in accepting their direct descent from the Dipnoans.

VII

THE TELEOSTOMES

ALL fishes not to be grouped among Sharks, Chimæ-roids or Lung-fishes, have been included in the fourth sub-class, Teleostomi. In this are to be merged the two time-honoured groups, Ganoids* and Teleosts, since it is now found that there are absolutely no structures of the one group that are not possessed by members of the other. The terms, therefore, "Ganoid" and "Teleost," must be used in a popular and convenient, rather than in an accurate sense ; the former to denote the "old-fashioned" Teleostome, with its rhombic bony body plates, and carti-laginous endoskeleton ; the latter, the modern "bony fish," with rounded, horn-like scales and its calcified endo-skeleton.

Teleostomes present so wide a range of variation that it becomes exceedingly difficult to include in a single definition their minor structural characters.

As a basis for the comparison of the Teleostomes, the characteristic structures of a single type, *e.g.* the Perch, might conveniently be taken. From these conditions, typical of a modern and highly specialized form, the simple structures of the ancient, more primitive, and ancestral Ganoids may afterward be readily understood.

---

* The term **Ganoid, as here** used (as far as p. 147), includes the **Crossopte-**rygians as well.

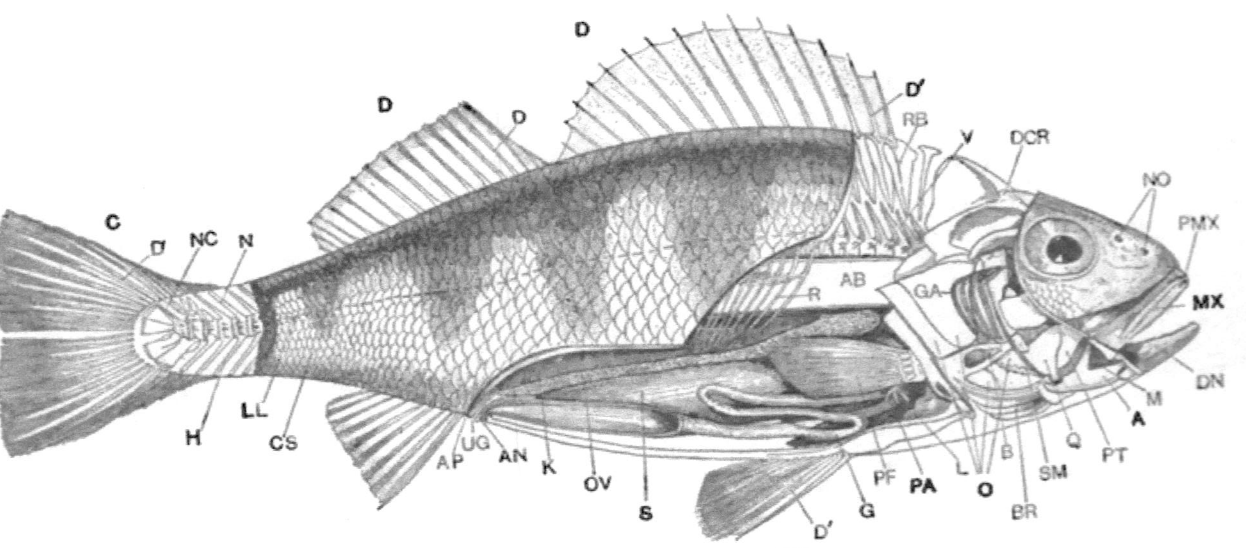

Fig. 145. — The general anatomy of a Teleost.

A. Anal fin. *AB.* Air-bladder. *AN.* Anus. *AP.* Abdominal pore. *B.* Bulbus arteriosus (conus is rudimentary). *BR.* Branchiostegal rays. *C.* Caudal fin. *CS.* Cycloidal scales. *D.* Dorsal fin. *D'.* Dermal supports of fins. *D.N.* Dentary bone. *DCR.* Dermal bones roofing cranium. *G.* Intestine. *GA.* Gill arches. *H.* Hæmal arch and spine. *K.* Kidney (mesonephros). *L.* Liver. *LL.* Lateral line. *N.* Neural arch and spine. *NO.* Anterior and posterior nares. *O.* Opercula. *OV.* Ovary. *PA.* Pancreas (pyloric appendices). *PF.* Pectoral fin. *PMX.* Premaxillæ. *PT.* Pterygoid. *Q.* Quadrate. *R.* Ribs, showing accessory supports. *RB.* Radial and basal fin supports. *S.* Stomach. *UG.* Urinogenital opening. *V.* Vertebra (centrum). *VF.* Ventral fin.

General *Anatomy*

In the Teleost (Fig. 145) the shortened and muscular body appears admirably adapted to the conditions of aquatic motion. Anteriorly it is broad and deep, its trunk muscles firmly attached to the bony prongs of the enlarged base of the skull, *DCR*, and to the solid, compact, calcified vertebræ, *V*, and their stout processes. The fish's tapering sides are encased in horn-like cycloidal scales, *CS*, a light, flexible armour, whose elements overlap, defending every point, and whose smooth and slime-coated surface provides the least possible resistance to motion. The fins, *D, C, A, PF, VF*, are light and strong, erectile and depressible; their rays are thin, narrow, spine-like, strong; they are entirely dermal, their cartilaginous supports sinking within the body wall, *RB*. The caudal is large and fan-shaped (homocercal), its crowded rays providing admirably its needed strength; its stout basal supports, compacted beneath the tip of the notochord, *NC*, show that its form is modified heterocercy. The pectoral fin, *PF*, has now taken its position high in the side of the body; its basi-radial supporting elements are reduced to a proximal row of a few small irregular plates.

The skeleton is completely calcified. The vertebral axis has undergone entire segmentation, the notochord persisting only between the cup-shaped faces of the centra; the vertebral arches and processes have merged with the centra, and those of the hinder region, *N, H*, with probably the basal fin supports as well. Ribs, *R*, usually with intersupporting processes, strengthen the walls of the visceral cavity, and represent calcifications of the myocommata, rather than transverse processes of the vertebræ. The skull is formed of compact bony elements; its carti-

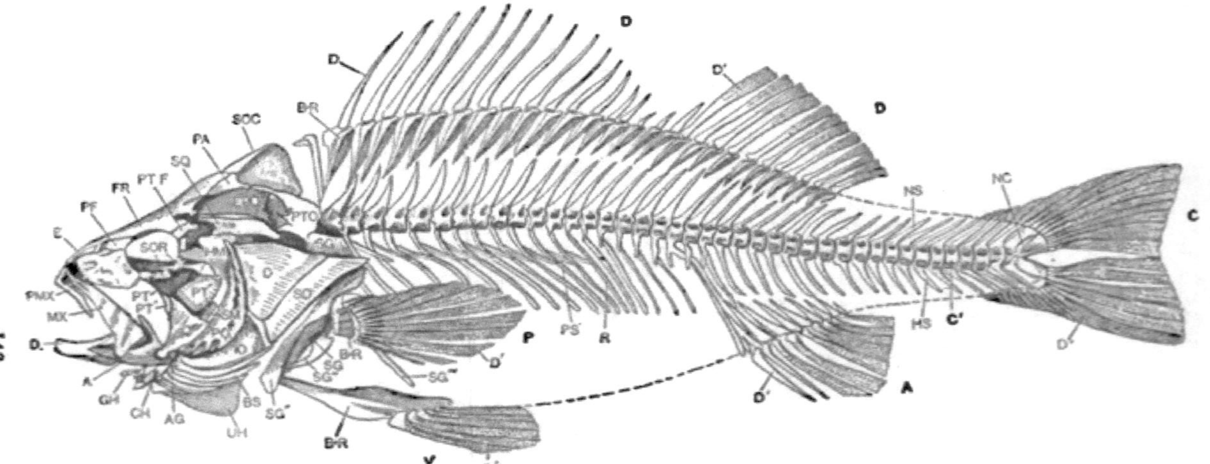

Fig. 146. — The skeleton of Teleostome. The Teleost, perch, *Perca fluviatilis*. × ⅓. (Drawn by Dr. A. GRAF, after ZITTEL.)
A. Anal fin. *AG.* Angular. *B+R.* Basal and radial cartilages. *BS.* Branchiostegal rays. *C.* Caudal fin. *CH.* Ceratohyal. *D.* Dentary. *D'.* Dermal rays of fins. *E.* Ethmoid. *EPO.* Epiotic. *FR.* Frontal. *GH.* Glossohyal (hypobranchial). *HM.* Hyomandibular. *HS.* Hæmal spine (and arch). *IO.* Interoperculum. *MX.* Maxillary. *NC.* Notochord. *NS.* Neural spine (and arch). *O.* Operculum. *P.* Pectoral fin. *PA.* Parietal. *PF.* Prefrontal. *PMX.* Premaxillary. *PO.* Preoperculum. *PS.* Supporting processes of ribs. *PT, PT',* and *PT''.* Ecto-, ento-, metapterygoid. *PTF.* Postfrontal. *PTO.* Pterotic. *Q.* Quadrate. *R.* Ribs. *SG.* Shoulder girdle. *SG'.* Dorsal process of *SG. SG''.* Outer rim of *SG. SG''', SG''''.* Posterior processes of *SG. SM.* Symplectic. *SO.* Suboperculum. *SOC.* Crest of supraoccipital. *SOR.* Suborbital ring. *SQ.* Squamosal. *UH.* Urohyal. *V.* Ventral fin.

laginous brain case is replaced by many definite osseous elements. The floor and roof of the skull, the face region, jaws, gill arches, and their protecting parts, are all encased by an elaborate series of membrane bones ; these, however, must be noted as deeply embedded in the body tissue, *DCR, DN, A, Q, PT, SM, BR, O.* The membrane bones of the jaw rim — maxillary, premaxillary, and dentary, *MX, PMX, DN* — bear teeth, and are especially characteristic of the Teleostomes ; those overlapping and protecting the gill arches (*GA*), *O, IO, PO, SO,* usually four in number, are also characteristic of the group. The skull is hyostylic.

As to the visceral parts. The gill arches, *GA*, are reduced in number, usually widely bent backward, and closely crowded together ; their gill filaments are enlarged and specialized. The heart lacks the arterial cone with its transverse series of valves ; in its place a stout bulbus, *B*, forms the base of the aorta. The digestive tract is tubular, long, and coiled ; its intestine, *G*, lacks a spiral valve, and terminates at the body surface, *AN*, not in a cloaca ; its glands include a series, often great in number, of pyloric cæca (pancreas). An air-bladder, *AB*, is present, which may, or may not, retain its communication with the gullet. The ovary, with its many small eggs, and the kidney, dorsal to it, have often a common external opening in a urino-genital papilla, *UG*, in either side of which abdominal pores may occur. The nervous system and sense organs (pp. 275, 277) have many peculiarities : the roof of the fore brain is non-nervous ; the nasal openings appear in the dorsal side of the head, *NO*, and are separate; the eye has specialized a vascular, nutritive structure, the *processus falciformis*, projecting from the region of the entrance of the optic nerve into the vitreous cavity of the capsule ; the optic nerves cross in passing to the eyes, but their fibres do not fuse.

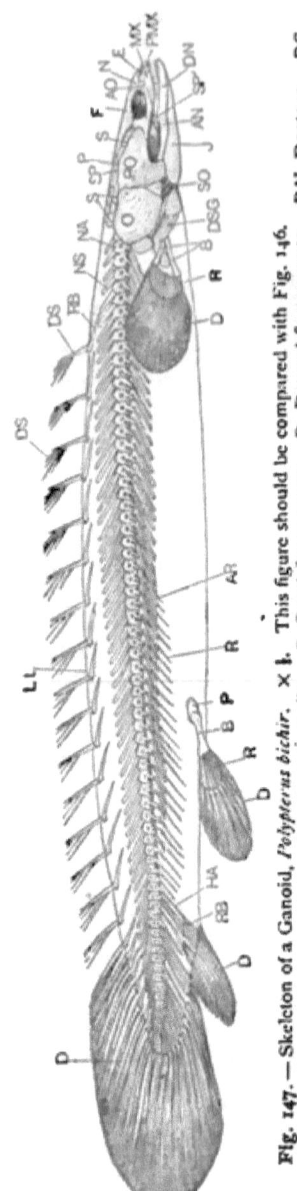

Fig. 147.—Skeleton of a Ganoid, *Polypterus bichir*. × ⅓. This figure should be compared with Fig. 146.
AN. Angular. *AO.* Anteorbital. *AR.* Accessory ribs. *B.* Basal fin supports. *D.* Dermal fin supports. *DN.* Dentary. *DS.* Dermal spine (a modified scale) of dorsal fin. *DS'.* Secondary dermal spines (radials, in part) of dorsal fin. *DSG.* Dermal scales functioning of outer shoulder girdle. *E.* Ethmoid. *F.* Frontal. *HA.* Hæmal arches. *J.* Jugular bones (scales). *LL.* Longitudinal ligament of dorsal fin. *MX.* Maxillary. *N.* Nasal. *NA.* Neural arch. *NS.* Neural spine. *O.* Operculum. *P.* Pelvic girdle. *PMX.* Premaxillary. *R.* Radial fin supports. *R'.* Rib (transverse process). *RB.* Radial and basal fin supports. *S.* Spiracular bones (scales). *SO.* Suboperculum. *SP.* Spiracle. *SP'.* Splenial.

Such in outline are the essential structures of a Teleost. They may now be briefly contrasted with the more important characters of the Ganoids.

In skeletal structures the Perch (Fig. 146) may be strikingly contrasted with the most nearly ancestral form of Ganoid (Fig. 147). In this, Polypterus (p. 148), the skeleton retains a semi-calcified condition. Its vertebral centra are practically separate from the arches; its ribs, *R*, are equivalent to the transverse processes; its accessory ribs, *AR*, to the "ribs" of Teleosts. The cartilaginous brain case is notably retained; the membrane, or dermal bones, of the head roof, as *F, P, SP, PO, O*, are clearly scale-like, with an enamelled surface, similar in character to those of **Dipterus.** The shoulder girdle **includes outer** dermal elements, *DSG.* The external parts of the unpaired fins are dermal; but their cartilaginous supports are retained, *RB*, even in the tail region. **The** caudal fin may be regarded as either diphycercal or heterocercal. **The** exposed parts of the paired **fins,** it **is especially interesting** to note, are only in part dermal; the two rows of **carti-** laginous supports are retained in **a** condition very similar to that of sharks, *R B*;* two of the basal elements of the pectoral fin, however, have retained the rod-like **form in** strengthening the front and hinder margin **of the fin.**

In visceral **structures the Ganoids** exhibit the following noteworthy **characters: a greater number** of gill arches; **a** spiracle; **a short and almost straight digestive** tube, with spiral valved **intestine;** a shark-like **pancreas;** an arterial cone, **with many** rows of **valves; a cellular air-** bladder, like that of **a** Dipnoan; primitive conditions in the urinogenital apparatus; **shark-like** characters in the **nervous** system and sense **organs; a** chiasma of **the optic** nerves, (pp. **260–279).**

Relationships and Descent

Johannes Müller, when separating Ganoids **from Tele-** osts, recognized clearly **even at** that early date (1844) that **the** majority of the structural differences of these forms were bridged over in exceptional instances; there were thus Teleosts with bony body plates, as well **as, it was** afterwards found, a Ganoid (*Amia,* **p. 163) with** herring-like cycloidal scales. But he believed that three structural characters of the Ganoids separated them constantly from all Teleosts, and warranted the **integrity of** the groups.

* Contrast Gegenbaur's view that this **fin represents the** simplest known condition of the archipterygium. *Ref.* on p. **248.**

L

These distinguishing characters were : —

I. A contractile arterial cone, containing rows of valves.

II. An intestinal spiral valve.

III. The interfusion (chiasma) of the optic nerve.

It was not until these differences were shown to be of little morphological importance that the two groups were merged in that of Teleostomi (Owen, 1866). Thus transitional characters in the arterial cone of *Butrinus* (p. 258) were discovered by Boas : the Teleost *Cheirocentrus* was found to present ganoidean intestinal characters'; and the optic chiasma, as Wiedersheim * demonstrated, could no longer be regarded as of taxonomic or morphological value.

The descent of the Teleostomes, like that of the other groups, has long been a matter of speculation. Their affinities with the Dipnoans are generally admitted (Günther, Gegenbaur, Haeckel, Smith Woodward). Rabl derives them directly from a selachian stem, regarding the Dipnoans as later evolved ganoidean forms. Beard, on the other hand, even goes so far as to entirely separate the Teleostome stem from that of the shark, lung-fish, and amphibian, deriving it with a close kinship to Petromyzonts, from the earliest vertebrates. Palæontology, however, has lately been giving rich contributions to this disputed problem, and there can at present be little doubt that the conditions in fossil fishes have demonstrated that in most ancient times Dipnoan and Teleostome were closely approximated. Although even in the earliest fossils they may be distinguished (*c.g.* by the arrangement of the head-roofing derm bones, v. p. 127), yet, as Smith Woodward has noted, forms occur too clearly transitional to indicate anything less

---

* One form of lizard was shown to possess a chiasma of the optic nerves; in its neighbouring genus the nerves were found to cross without fusion.

than genetic kinship. The Crossopterygian, whose ancient structure is well known, may well have been **derived from** an ancestor common to the Ctenodont (Dipnoan) **and** Holoptychian (Fig. 153); so that the gradual **nearing of the** Teleostome stem to that more fixed, **of the Dipnoan, is a** strong suggestion as to **its** derivation. The later **descent** of the Ganoids from **an** ancestor closely akin to, if not identical with the Crossopterygian, is usually conceded. Teleosts first occurring in Cretaceous are by evidence of fossils the almost undoubted survivors **of an** extensive group **of** transitional Mesozoic Ganoids **(p.** 165). But whether all Teleosts are **to be deduced from a single** ganoidean phylum can **at present hardly be established.** Thus catfishes, or Siluroids, appear in many **structural** regards closely **akin** to the sturgeon (p. 160); **but as their** fossil remains are lacking before the Eocene—**when, how-** ever, they appear **to have** been in every **way as** highly evolved as in recent forms—little clue has been given **to** their descent.

Teleostomes may, **in** the **present connection, be briefly** characterized under their two principal subdivisions.

I. Crossopterygian, **the more** archaic **group, uniting** characters of shark, lung-fish, **and Ganoid, retaining the** ancient cartilaginous fin bases, radials, **and** basals **in their** lobate fins; in some forms (Holoptychius, **Fig.** 153), **the** concrescence of the basal parts of unpaired **fins passing** through the same evolution as those of paired fins. Represented in the surviving Polypterus (**"Bichir" of** the White Nile, Fig. 148), and in **the slender** Polypteroid Calamoichthys (of Calabar), and in the extinct Holoptych- ius, Undina, Diplurus, and Cœlacanthus.

II. Actinopterygian, **the spine-finned** Teleostomes. Fins supported by dermal rays; ancient fin support greatly

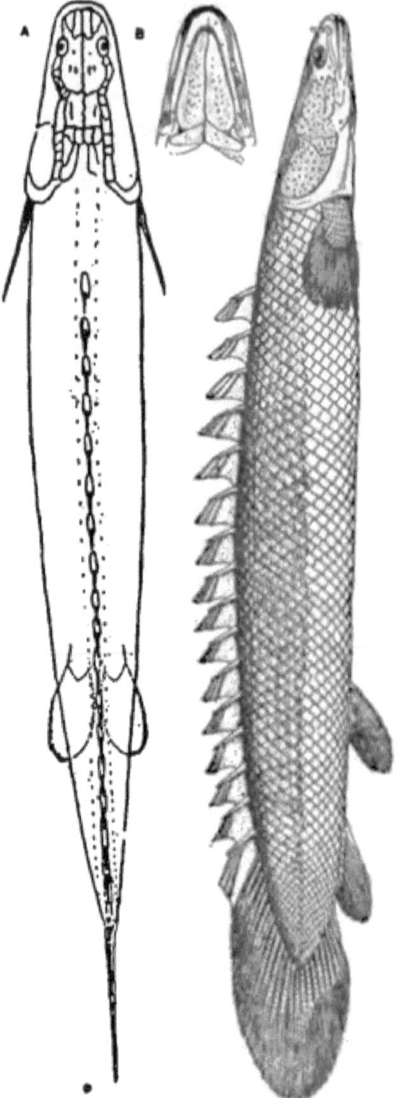

Fig. 148.—The Nile bichir, *Polypterus bichir.* × ¼. White Nile. (Modified after L. AGASSIZ.)

A. Dorsal aspect. *B.* View of throat region, showing jugular (gular) plates and ventral elements of the dermal shoulder girdle.

reduced, implanted within body wall. Includes *Chondrosteans* ("Ganoids") and *Teleocephali* ("Teleosts").

I. CROSSOPTERYGIANS

The CROSSOPTERYGIANS, as palæontology has demonstrated, are the most ancient Teleostomes. In their structural characters — especially in the fins, skeleton, nervous system — they are clearly to be separated from the neighbouring Ganoids. And their transitional characters have not as yet been clearly demonstrated.

Polypterus (Figs. 148, *A, B,* 149) and its kindred genus, *Calamoichthys* (Fig. 150), stand alone as the survivors of the Crossopterygian group. They have diverged but little from their Devonian kindred, and demonstrate in the most interesting way the persistent survival of fishes. From their isolated position, these recent forms become of extreme interest to the morphologist, and from the side of their development, when this comes to be studied, they are expected to throw the greatest light on the relations of the primitive Teleostome to the sharks and Dipnoans, on the one hand, and to the Ganoids on the other.

Polypterus * presents the exoskeletal characters of the ancient Crossopterygians, and the typical conditions of their lobate pectoral fins ; the dermal plates of its head region are tuberculate as in Dipnoans, but, unlike these, their arrangement, as in all Teleostomes, is dis-

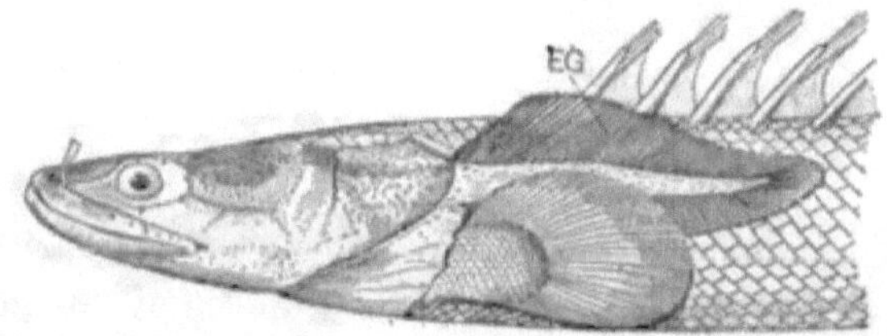

Fig. 149.— *Polypterus lapradei.* (After STEINDACHNER.)   Head region of well-grown larva showing **external gill**, *E G.*

tinctly paired, *i.e.* "ethmoids," frontals, parietals, occipitals (Fig. 148 *A*), including a pair of gular plates in the throat region, *B.*†   Among the structures peculiar to the

* Polypterus occurs in the Nile, but is rarely taken below the Cataract. It was noted, however, from near Cairo in the *Description d'Egypte*, and a specimen in the possession of Professor Innes of the College of Medicine, Cairo, was taken near Boûlak a few years ago. It is known by the Arabs near Assuan, and is here occasionally taken in the fykes at the beginning of the flooding-season. The remarkable series of Polypterus in the Vienna collection was collected in the White Nile, although some of these specimens, Dr. Steindachner has stated personally to the writer, were taken in Middle Egypt. It seems evident to the writer, from the results of his collecting-trip from Cairo to Assuan, April and May, 1892, that abundant material of Polypterus is not readily secured below the Second Cataract. Until, therefore, the interior of Egypt is made more accessible to foreigners, developmental stages can hardly be hoped for.

† As in some of the fossil lung-fishes.

recent forms may be included the fringing dorsal fin, the tubular nasal opening (Fig. 149), and an external gill in Polypterus (Steindachner), *EG*, in the late larval stages.

Calamoichthys is unquestionably a divergent member of the stem of Polypterus; its form, becoming elongated, has acquired a general undulatory movement; the paired fins have accordingly diminished in relative size, the ventral fins finally disappearing.

Little is known of either the living or breeding habits of Crossopterygians: in these they might naturally be expected to resemble the Ganoids.

Fossil Crossopterygians

A number of the fossil kindred of Polypterus are shown in the succeeding figures (Figs. 151–156 *A*).

Gyroptychius and *Osteolepis*, Devonian genera (Figs. 151, 152), are certainly most nearly in the ancestral line of the recent forms. Like many sharks and fossil Dipnoans, they present a heterocercal tail, a single anal fin, and a pair of dorsals. The pectoral fin of Osteolepis is becoming a typical archipterygium.

Holoptychius, another Devonian form (Fig. 153), approaches even more closely the dipnoan types: the scales are cycloidal; its paired fins are distinctly archipterygial; and the caudal region, reduced in length, is becoming metamorphosed into the typical diphycercal form by the tendency of the second dorsal and anal fin to coalesce with

Fig. 150.—*Calamoichthys calabaricus.* × ⅓. Senegambia.

the caudal. In these forms a number of paired gular plates may occur.

In a closely related genus, *Eusthenopteron*, also of

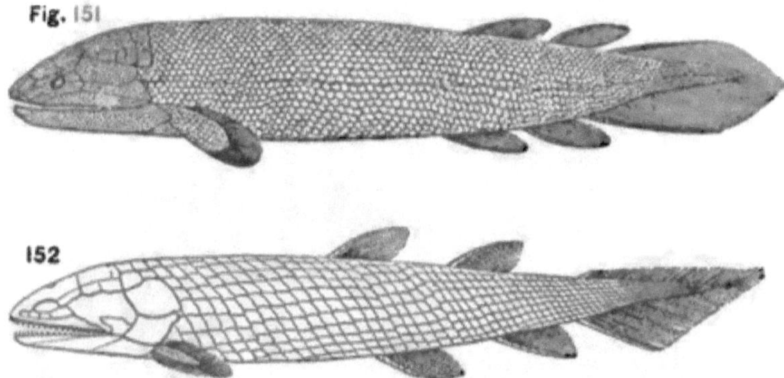

Fig. 151.— *Gyroptychius*. × 1. Old Red Sandstone, Scotland. (After SMITH WOODWARD.)

Fig. 152.— *Osteolepis*. × 1. Old Red Sandstone, Scotland. (Restoration from SMITH WOODWARD, after PANDER.)

Devonian age (Fig. 154, *A*, *B*), the structure of the basal parts of the unpaired fins is exceedingly interesting; the radial supports are unfused, while the basals, merged in a

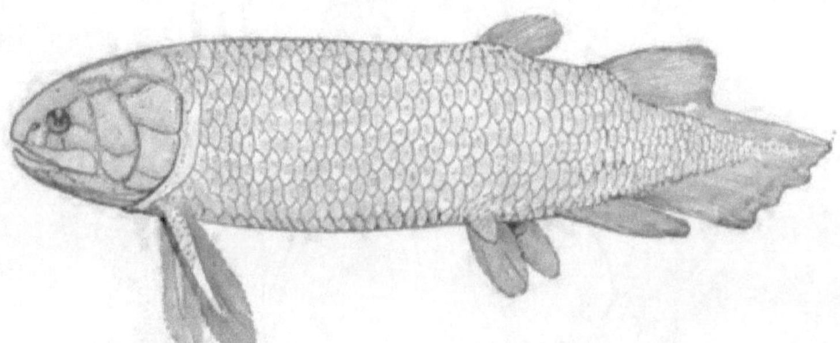

Fig. 153.— *Holoptychius andersoni*. Old Red Sandstone, Scotland.

single plate, have come into especial relation with the axial skeleton; the subsequent stage of their differentia-

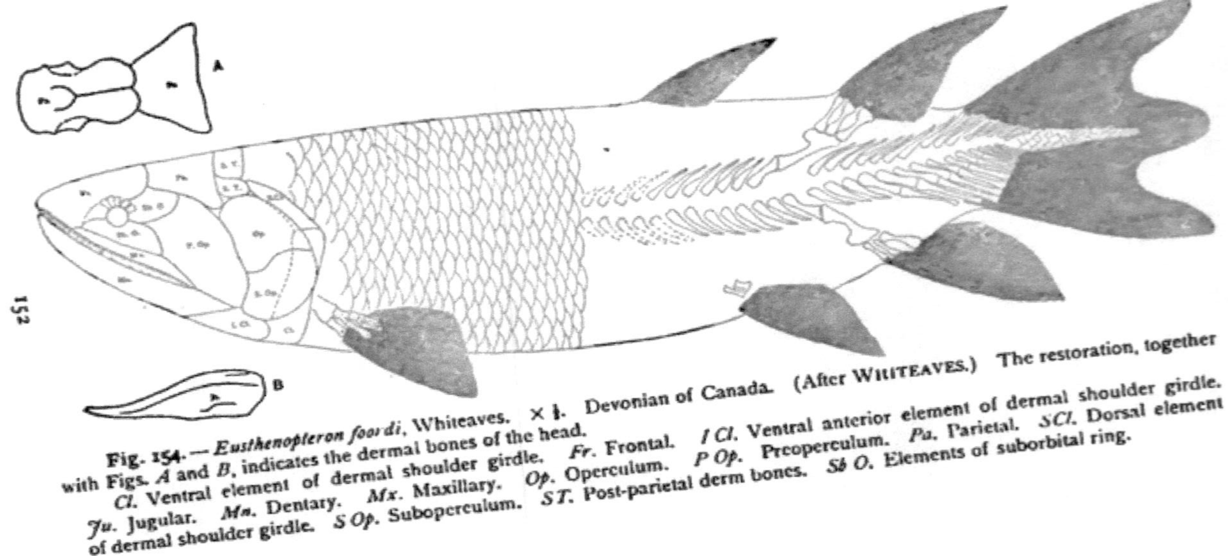

Fig. 154.—*Eusthenopteron foordi*, Whiteaves. × ⅓. Devonian of Canada. (After WHITEAVES.) The restoration, together with Figs. *A* and *B*, indicates the dermal bones of the head. *Fr.* Frontal. *I Cl.* Ventral anterior element of dermal shoulder girdle. *Cl.* Ventral element of dermal shoulder girdle. *Op.* Operculum. *P Op.* Preoperculum. *Pa.* Parietal. *SCl.* Dorsal element of dermal shoulder girdle. *S Op.* Suboperculum. *ST.* Post-parietal derm bones. *Sb O.* Elements of suborbital ring. *Ju.* Jugular. *Mn.* Dentary. *Mx.* Maxillary.

152

tion has been noticed in Fig. 43. The condition of the caudal fin of Eusthenopteron is also worthy of note ; the tip of the notochord is retained although the functional portion of the fin is derived from the more anterior body region. The vertebral arches are here clearly suggestive of the conditions of the Dipnoan.

Cœlacanthus, common in the Coal Measures (Fig. 155), is the most specialized of the Crossopterygians ; it has retained all of the archaic structures of its kindred, yet has concealed them under the outward appearance of a recent bony fish ; the general contours of its head, trunk, scales and fins resemble strikingly those of a dace or

Fig. 155. — *Cœlacanthus elegans*, Newb. × ⅓. Coal Measures, Ohio.
A. Position of calcified swim-bladder.

chub ; but on closer view the paired fins are found to be archipterygial, the scales enamelled and sculptured, the true caudal fin the degenerate stump of the notochord ; the functional caudal has been formed of the enlarged fin rays of the dorsal and anal region. Traces of a calcified air-bladder, *A*, are often preserved.

Diplurus and a closely related genus, *Undina* (Figs 156, 156 *A*), may finally be noted among the highly evolved Crossopterygians. They appear in the Mesozoic when the majority of their kindred have disappeared ; they have assumed peculiar characters and have apparently reached the point of differentiation when they shortly become extinct.

Diplurus has become excessively shortened in its body length; the head is of relatively enormous size; its derm bones are squamous, and appear to have been deeply implanted in the integument; teeth have disappeared;

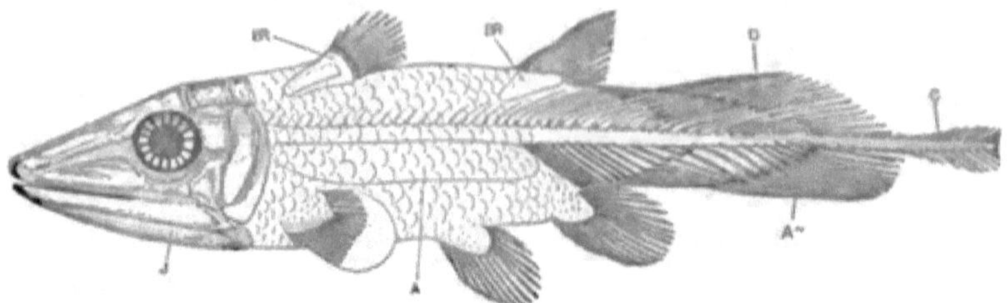

Fig. 156. — *Diplurus longicaudatus*, Newb. × ⅓. Triassic, Boonton, N.J. *A*. Position of calcified swim-bladder. *A″*. Second anal fin (now the ventral portion of the functional caudal). *BR*. Radial and basal fin supports. *C*. Caudal fin (degenerate). *D*. Hindmost dorsal fin (now the dorsal portion of the functional caudal). *J*. Jugular.

scales have become exceedingly thin and are rarely preserved. Fin structures are apparently of a degenerate character; their cartilaginous bases, when showing, appear

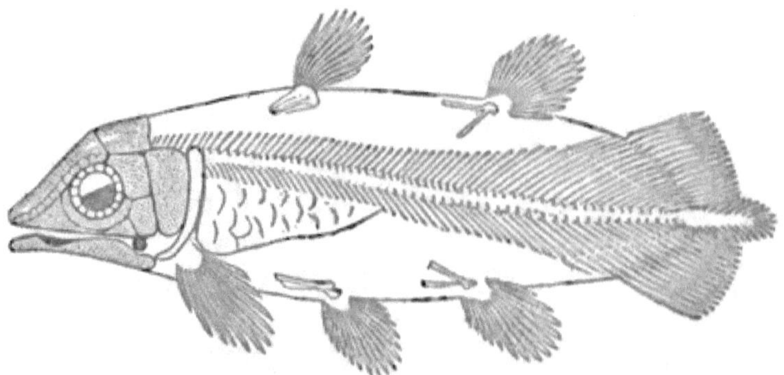

Fig. 156 A. — *Undina gulo*, Egert. × ¼. Lower Lias of Lyme Regis. (Restoration after Smith Woodward.)

to have become reduced to single plates, as *BR*; the caudal is the elongate tip of the vertebral axis; the functional caudal, now elongate and diphycercal, is formed by dorsal and anal elements, *D*, A'', as in Cœlacanthus. The boundary line of the calcified air-bladder, *A*, is often preserved.

II. ACTINOPTERYGIANS

A. *Chondrosteans* (*Ganoids*). Ganoids agree with the Crossopterygians in their exoskeletal characters, although usually lacking in gular plates. The most important differences between these groups have been reduced to those of fin structures; the Ganoids have no longer the lobate form of the paired fins; their basal fin supports have become greatly reduced and are usually represented by a single row of a few metamorphosed elements in the

Fig. 157.—The short-nosed gar-pike, *Lepidosteus platystomus*, Raf. × ⅓. Mississippi basin. (After GOODE in U. S. F. C.)

most proximal region of the fin. The transitional stages — if they exist — between the lobate and the monoserial fins have not as yet been demonstrated.

Fossil Forms

From the middle of the Palæozoic period to the end of the Mesozoic there seems to have been a culminating time of forms like the still existing Gar-pike (Fig. 157); their fossils are generally the most numerous, and, on account partly of their strong body armouring of interlocking rhombic plates, the most perfectly preserved of fossil fishes. They usually exhibit the structural characters

which Lepidosteus has retained, while diverging widely on
all sides in matters of shape, size, special dentition, and

Fig. 158.—*Elonichthys* (*Rhabdolepis*) *macropterus* (Giebel), Bronn. × ¼.
(After L. AGASSIZ.) Lower Permian, Rhenish Prussia.

features of the body armouring,—characters, apparently,
of minor morphological importance. But a few of the characteristic types of the early Ganoids can be noted in the
present connection. Some of the more important have
been figured in Figs. 158–164.

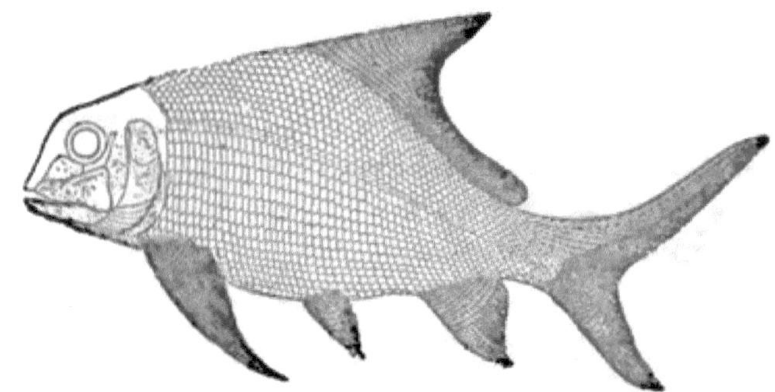

Fig. 159.—*Eurynotus crenatus*, Agassiz. × ¼. (After TRAQUAIR.) Calcif-
erous Limestone, Scotland.

Thus *Elonichthys* (Fig. 158) was a form which had
evolved a small size and narrow sculptured body plates;

Eurynotus (Fig. 159) had attained a great depth of body and prominent dorsal fin ; *Cheirodus* (Fig. 160) was distinctly flattened ; *Semionotus* (Fig. 161) was small, with

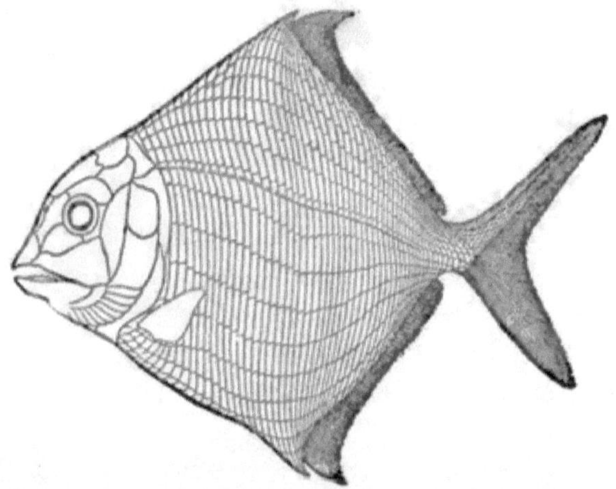

Fig. 160.— *Cheirodus granulosus*, Young. × ¼. Coal Measures, Scotland. (After TRAQUAIR.)

elaborate fin conditions ; *Aspidorhynchus* (Fig. 162) had a remarkable pointed snout and a reduced number of body

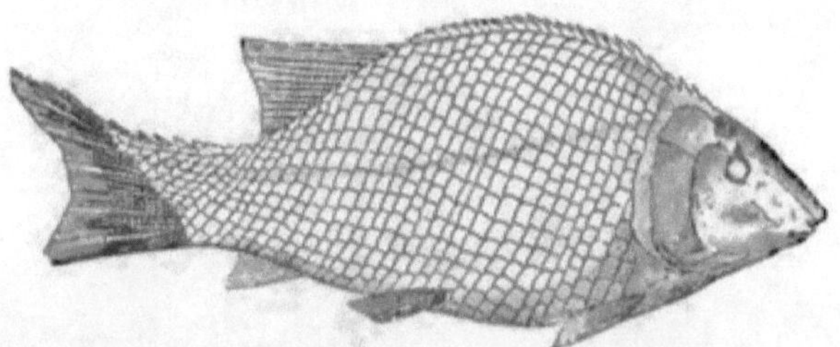

Fig. 161.— *Semionotus kapffi*, Fraas. × ⅓. (From ZITTEL, after FRAAS.) Keuper, Stuttgart.

plates ; *Microdon* (Fig. 163), flattened like Cheirodus, had evolved an admirable series of crushing teeth (-Pycnodont).

And, finally, is to be mentioned *Palæoniscus* (Fig. 164), a form whose abundance, numerous species, and long sur-

vival (Palæozoic-Mesozoic) have made it the most widely
known of fossil fishes.   Of all extinct Ganoids there

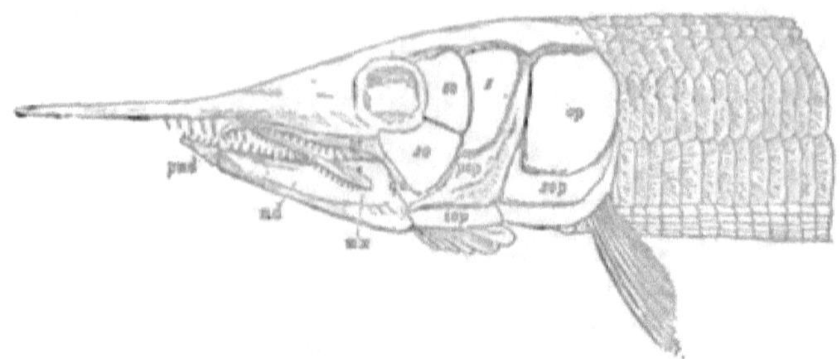

Fig. 162. — *Aspidorhynchus acutirostris*, Agassiz.   × ⅓.   (After ZITTEL.)   Jura, Solenhofen.

appears to attach to Palæoniscus the greatest morphological
interest ; on the one hand, it seems closely akin to the

Fig. 163. — *Microdon wagneri*, Thiollière.   × ¼.   (From ZITTEL, after THIOL-
LIÈRE.)   Jura, Cerin.

recent gars, and, on the other, even as evidently to the sturgeons ; of all fossil kindred of these living forms, it seems most nearly in the ancestral line.

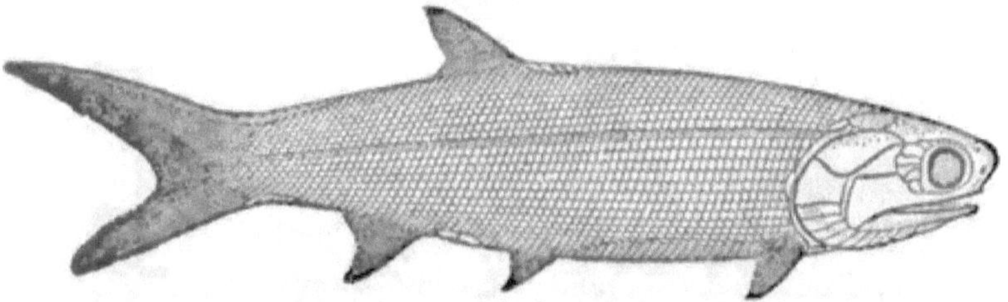

Fig. 164.—*Palæoniscus macropomus*, Agassiz. × ?. (After restoration of TRAQUAIR.) Upper Permian.

Ganoids certainly outrank the Crossopterygians in the number and variety of their ancient forms. Their few living representatives give but little idea of the importance of the group, and can suggest but faintly the lines of its evolution.

Living Types

The recent Ganoids include the Gar-pike, the Sturgeons, and Amia. The first is of especial interest in connecting the group most closely with the Crossopterygians, the last as best illustrating the intermediate stage between the Ganoids and Teleosts.

The Gar-pike, *Lepidosteus* (Fig. 157), resembles Polypterus in many characters of skeleton and dermal defences. It is a form not uncommon in the fresh waters of North America, and is especially abundant in the Mississippi, Great Lakes, and rivers of the Southern States. In South Carolina the writer has known the gar-pikes to occur in such numbers that they would fill the shad nets, and for many days render this fishery impracticable. They sometimes attain a length of six feet, and are said to become

as aggressive as sharks. They are remarkably tenacious
of life, and their complete armouring of dermal plates
renders them practically invulnerable.

In development Lepidosteus has apparently more primitive features than Acipenser (v., p. 207 ; also *Jour. of
Morph.* XI, No. 1).

Of all recent Ganoids, Lepidosteus must certainly be
looked upon as retaining most perfectly the structural
characters of the most abundant and probably the most
generalized Palæozoic and Mesozoic forms. Its genus, it
is true, is not known to occur earlier than the Eocene, but
its structures — scales, fins, labyrinthine teeth and partially
calcified skeleton — are known to have been possessed,

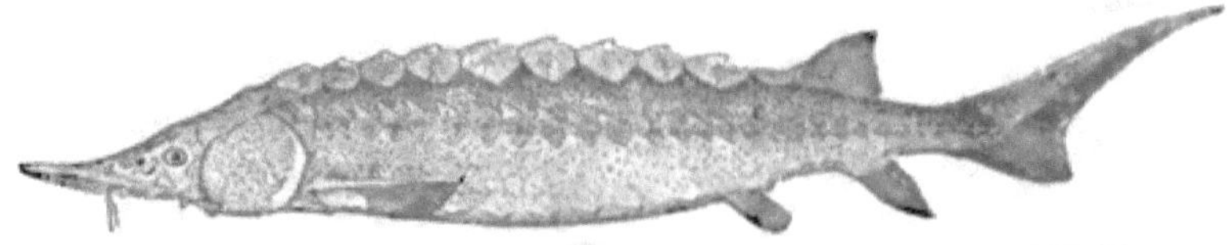

Fig. 165. — The sturgeon, *Acipenser sturio*, L. × ₇. Streams entering North
Atlantic. (After GOODE in U. S. F. C.)

even in their details, by a number of the older genera and
families.

The Sturgeons, *Acipenser, Scaphirhynchus, Psephurus,
Polyodon,* must in many ways be looked upon as of a highly
adaptive or even retrogressive character. There is strong
evidence that in their descent a large proportion, and, in
cases, all of their dermal armouring has been lost, and that
their cusp-like ancestral teeth have either disappeared or
are retained in a rudimentary condition.

The interrelationships of the four surviving forms of
sturgeons have not as yet been definitely suggested ; transitional fossil forms have thus far been lacking, and the
relative importance of the different structures in the recent

genera cannot, therefore, be determined for purpose of comparison.

The genus of the common sturgeon, *Acipenser*, is the most completely studied of the recent forms. It includes twenty or more "species," varying in length from one (*A. brevirostris*, of the Eastern United States) to ten yards (*A. huso*, of Russia), and is altogether one of the most valuable food-fishes of the rivers, lakes, and coasts of the northern hemisphere. It is a sluggish, bottom-feeding fish, common in muddy streams. Its broad and pointed snout, sensory barbels, and greatly protractile jaws are the most striking differences from the Palæoniscoid; its dermal

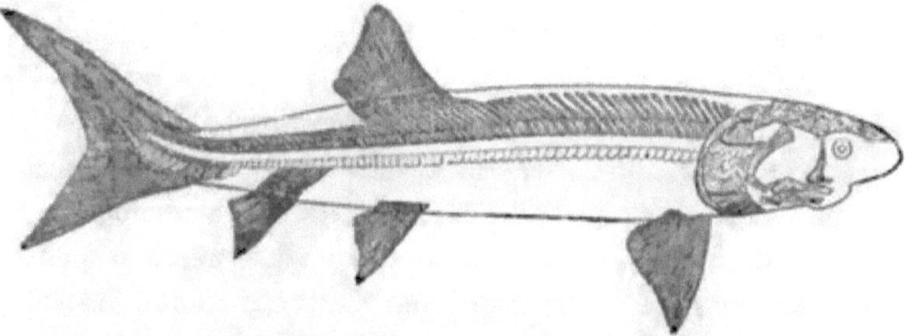

Fig. 165 A. — *Chondrosteus acipenseroides.* × ¼. From Lias of Lyme Regis. (Restoration of skeleton after SMITH WOODWARD.)

armouring has become reduced to the five longitudinal bands of body plates,* but is more perfect in the tail region; its skeleton retains an entirely cartilaginous condition. In its larval stage conical teeth are known to be present, and the entire series of dermal plates are much larger in relative size.

A figure of *Chondrosteus*, a Liassic sturgeon, may here

* It is interesting to note that in Palæoniscoids there is sometimes a noticeable tendency for the five rows of plates, dorsal, and the paired lateral and ventral, to increase in size, suggesting the first steps in the origin of the derm plates of Acipenser.

M

parenthetically (Fig. 165 *A*) be inserted; it is of especial interest as suggesting an approximation of the type of the modern sturgeon to that of the Palæoniscoid; its snout is shorter than in Acipenser; its jaws larger, and apparently less protrusible; its dermal plates of the head region, including the branchiostegals, are clearly of the ancient pattern, and the fins, fin supports, and vertebral characters,

Fig. 166.—The shovel-nose sturgeon, *Scaphirhynchus platyrhynchus* (Raf.), Gill. × ¼. Mississippi basin. (After GOODE in U. S. F. C.)

together with the general small size of the fish, suggest intermediate conditions.

Of the remaining sturgeons, the shovel-nose, *Scaphirhynchus* (Fig. 166), of the Mississippi and of Central Asia, seems to possess the closest relations to Acipenser; although it is apparently a more modified form, on account of its elongate body shape and flattened snout, it still retains many interesting and archaic features. Among

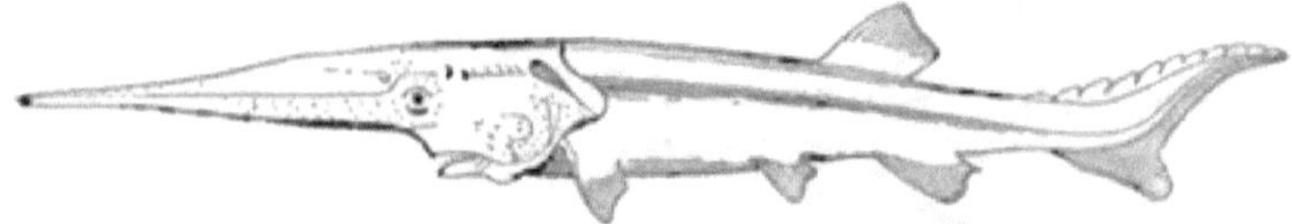

Fig. 166 A.—*Psephurus gladius*, Gün. × ¼. Rivers of China. (After GÜNTHER.)

these it includes the most complete dermal armouring of recent forms, its hinder body region being entirely encased.

Psephurus (Fig. 166 *A*), of the Chinese rivers, and *Polyodon*, or Spatularia (Fig. 165 *B*), of the Mississippi, are the other forms of living sturgeons. Their greatly elongate snouts, giving them the popular names of Spoonbills, Paddle-fish, Spear-fish, are among the most remarkable sensory appendages of fishes. They have been but little studied, and their relations to Acipenser have never been satisfactorily determined. They have certainly many feat-

Fig. 166 B. — The spoon-bill sturgeon or paddle-fish, *Polyodon spatula* (Walb.), J. and G. × ⅛. Ventral and side view. Mississippi basin. (After GOODE.)

ures in skeletal parts, fin structures, lateral line organs, jaws, teeth, which can only be looked upon as of primitive character; on the other hand, their highly specialized rostrum, degenerate opercula, and want of dermal amouring would suggest an early divergence from the main stem of the sturgeons. To the writer, Psephurus seems the more generalized of these peculiar forms.

Fig. 167. — The bowfin, *Amia calva*, L. × ⅛. (After GOODE in U. S. F. C.) Central and Eastern United States.

Amia calva (Fig. 167) is the last of the recent Ganoids to be noted. Its distribution corresponds closely with that of the gar-pike; it is a common form, worthless as a food-fish, but deemed worthy of a host of local names, as: Bowfin, Grindle, Dog-fish, Mud-fish, Sawyer, Joseph Grindle, Lawyer-fish. Its interest, as already suggested, is in its close kinship to the Teleosts on the one hand, and to the sturgeons and gars on the other. Its cycloidal scales, its fin structure, and calcified skeleton seemed of so modern a character, that it was long included among the members of the herring group; only after a closer examination did its primitive structures become apparent. It is one of the few Ganoids which possess a gular plate (Fig. 168, *jug*). Like that of Lepidosteus, its air-bladder is cellular, and of respiratory value (Wilder).

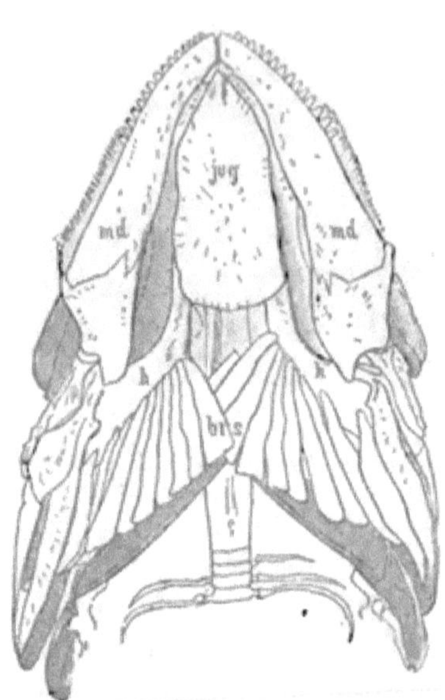

Fig. 168. — *Amia.* Ventral view of jaw region. × 1. (After ZITTEL.).
brs. Branchiostegal rays. *h.* Cerato-hyal. *jug.* Jugular plate. *md.* Mandible.

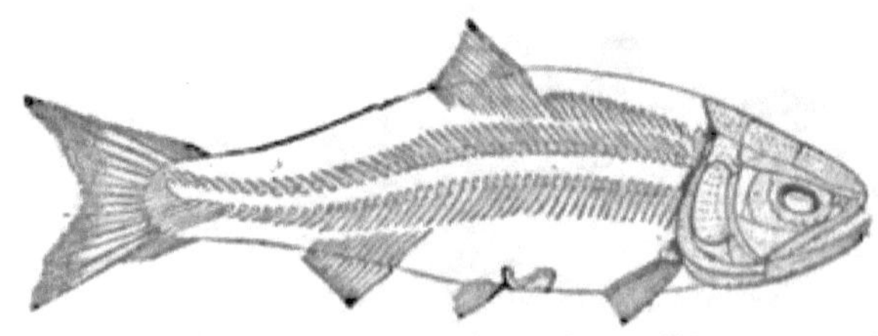

Fig. 169. — *Caturus furcatus.* × ⅓. (From SMITH WOODWARD, after AGASSIZ.) Lithographic stone (Upper White Jura), Solenhofen.

The relations of Amia become of especial interest, in view of the number and range of its fossil kindred. Its

group is known to have attained its prominence at a later
geological time than the other Ganoids; it is doubtless
derived, more or less directly, from the main ganoidean
stem. Three of the more typical Mesozoic forms are
shown in Figs. 169, 170, 171, in *Caturus*, *Leptolepis*, and

Fig. 170.—*Leptolepis sprattiformis.* × ⅓. (From SMITH WOODWARD.) Lithographic stone, Solenhofen.

Megalurus. To these amioid forms the ancestry of the
(majority of the) Teleosts is reasonably to be traced.

A general scheme of the phylogeny of the Teleostomes
is suggested on the adjoining page (Fig. 171 *A*).

B. *Teleocephali* (*Teleosts.*)   This group, popularly known
as that of the bony fishes, or Teleosts, includes as great
a proportion perhaps as 95 per cent of the kinds of fishes

Fig. 171.—*Megalurus elegantissimus,* Wagner. × ⅓. (After ZITTEL.) Jura,
Solenhofen.

living at the present time. The immense number of their
genera and species is doubtless suggestive of the form
changes which occurred during the flowering periods of
the sharks, chimæroids, or lung-fishes.

Teleosts have diverged most widely of all fishes from

what seem to have been their primitive structural conditions. Their skeleton has become highly calcified, its elements multiplying, fusing, and specializing. The notochord
has practically disappeared, owing to the complete formation
of bony vertebræ. The derm bones of the head, which in

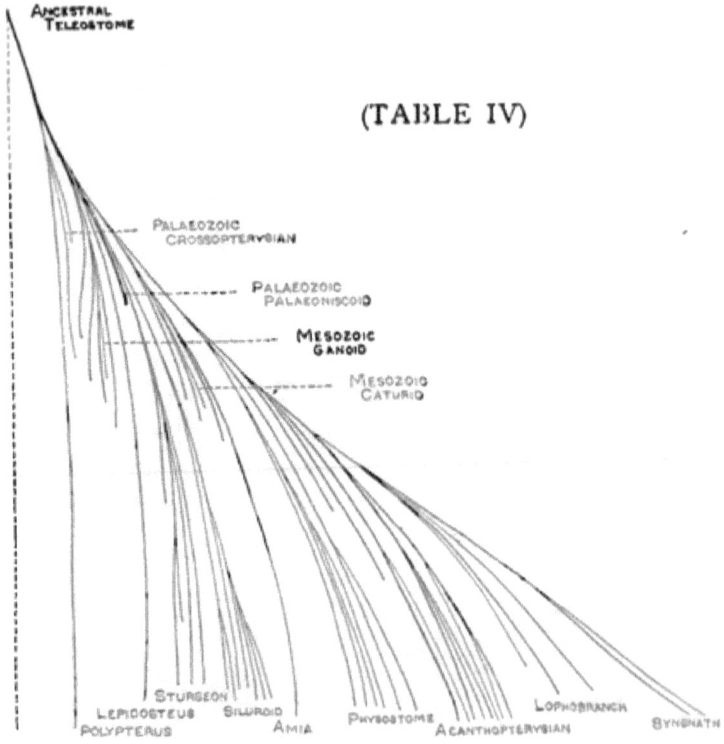

Fig. 171 A. — The Phylogeny of the Teleostomes.

the ancestral Ganoid were at the surface, enamel-coated,*
are now deep-seated in the head, resembling true cartilage
bones; their surfaces are usually deeply furrowed or ridged,

---

* The enamel of Ganoid plates (ganoine) appears to be derived from the
underlying bony tissue, not deposited by the overlying epidermis (enamel
organ).

and their character is often **squamous.** **Scales are** widely specialized, thin, horn-like, ornate, overlapping their outer margins, their inner rims set deeply but loosely in dermal pockets (Fig. 31). Fins are dermal structures, their **ancient** basal supports hardly to be distinguished; the primitive tail structure is so masked by clustered and fused **skeletal** elements that its heterocercy is scarcely apparent. **In** short, the most widely modified conditions can be shown to exist in Teleosts in almost every structural **character,** as in gills, teeth, opercula, circulatory and **urinogenital** organs, sensory structures, and **nervous system.** **They** have evidently been competing **keenly in the** struggle for **survival,** for in every detail of **form or structure** the most **varied** conditions **exist.** **In addition to these structural** adaptations of **Teleosts, changes in coloration have been** rendered possible by the transparency of their scales; **and** in their different families these changes **have** taken place often with striking **results:** adaptive coloration, brilliant, **dull,** mottled, inconspicuous, **occurs with a range of variation which is** not surpassed even by **the colours of birds.**

It is not remarkable, **therefore, that** members of the different groups **of** Teleosts should **often parallel each** other in structural likenesses, when placed **under the same** environmental conditions. Each organ, in fact, may become a centre **of** variation, and confuse the **line** of the descent **of the** minor groups; for the keenest judgment cannot select of all these varying structures those which can definitely be made the standards of general comparison. Environment, like a mould, has impressed itself upon forms genetically remote, and in the end has placed them side by side, apparently closely **akin,** similar in form and structure.

A striking instance of changes due **to** environment is

well known in the case of *Deep-sea Fishes*, in their acquiring a characteristic shape under the conditions of abyssal
life. The head region of these forms becomes greatly
exaggerated in size, and the trunk tapers suddenly away
toward the tip of the pointed tail. The tissues become
extremely modified, soft, porous, delicate, often transparent ; skeletal parts are deficient in lime, and loosely
articulated. Many organs are retained in curiously undeveloped or aborted conditions ; the vertebral axis is noto-

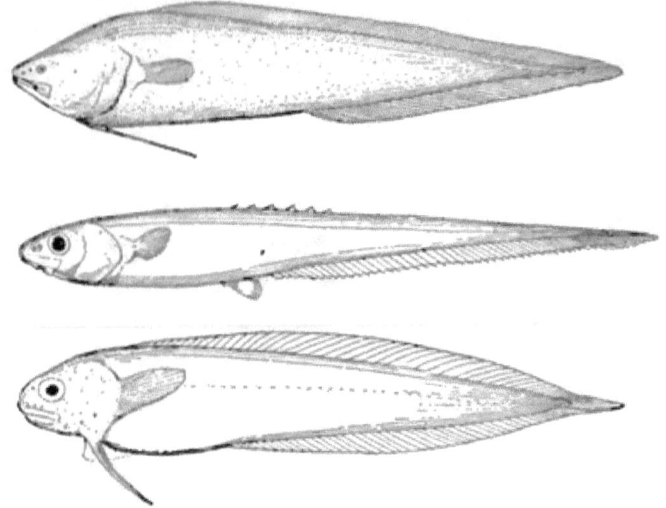

Figs. 172-174. — Deep-sea fishes. (After GÜNTHER.) 172. *Paraliparis bathy-
bius.* 640 fathoms. 173. *Bathyonus compressus.* 1400 fathoms. 174. *Notacanthus
sexspinis.* 1800 fathoms.

chordal ; gill arches, as many as six (?) in number, may open
freely to the surface, never enclosed by opercula ; sensory
canals remain as open grooves as in the most generalized
fishes ; paired fins are retained either in an undeveloped
condition or are not produced at all. Absence of light has
been not without its effects ; body colours are usually dark
and meaningless ; while, on the other hand, when eyes still

occur, a widely modified series of integumentary phosphorescent organs are **often** evolved as **lures by predatory** forms. It is evident, **in** the case of deep-sea fishes, that **the** simple condition of their structures does not separate **them** widely **in point** of **descent from more** specially evolved Teleosts. Intermediate **forms,** occurring **in shallower** water, often connect **them** clearly with different, and widely distinct, **groups of bony** fishes. **In** this way the

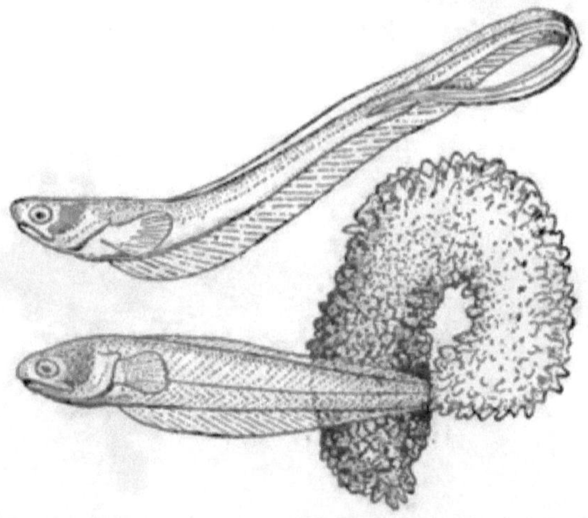

Fig. 175. — *Fierasfer acus*, Kaup. × ⅓. (After EMERY.) Commensal of sea-cucumber in southern waters.

forms which are **shown in Figs. 172, 173, 174 are** severally connected **with** the cottid, **the cod and the salmon, al**though the striking similarity **of their outward structures** would **naturally** lead **one to regard them as far more intimately** related.

Another interesting instance of the modification of a fish's form by its living conditions has often been noted in the case of *Fierasfer* (Fig. 175). **This small Teleost lives as a commensal in the** branchial **chamber of the sea-cucum-**

ber, and from its peculiar life habit retains permanently
a number of its embryonic characters ; it has thus its
elongated larval form, a functional pronephros, a notochordal skeleton and immature fin conditions (Emery,
Ref. p. 249).

To what degree the structures of fishes may be varied
by artificial selection is an interesting question, but one
that has as yet received little attention even from those
who have made artificialization an especial study.   In the
instance of the *Goldfish* it is well known how wide a

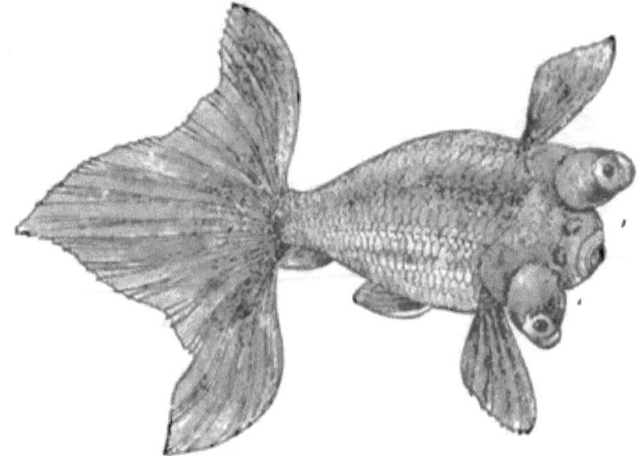

Fig. 176. — Goldfish, *Carassius auratus* ("Telescope" variety). × 1. (After
GÜNTHER.) Japan.

variation has been produced in colour, size, and proportions.
Fin structures are elaborately developed, long, drooping,
lace-like, often to a degree which must render progression
both slow and difficult.   Even the eyes have been made
to become large and protruding (Telescope-fish, Fig. 176).
In carp the variation in scale character, due to artificialization, is also to be mentioned.   It is natural, perhaps, that
artificial selection has been most successfully practised

among these forms which compete most actively for survival.

To conclude the present chapter, several forms of Teleosts may be briefly discussed as especially characteristic of the group, namely the catfish, Mormyrus, eel, perch, cod, flounder, porcupine-fish, sea-horse.

The catfish, representing the *Siluroids*, has, as already noted, many structural affinities to the sturgeon, and is, perhaps, a direct descendant of some early type of Mesozoic Palæoniscoid. It is a representative of a large and wide-spread family, usually of river fishes. Its habits are slug-

Fig. 177.—The bull-head (catfish), *Amiurus melas* (Raf.), Jord. and Copeland. × ¼. (After GOODE in U. S. F. C.) Eastern North America.

gish and mud-loving. Its trunk is heavy, rounded, and without Teleostean scales; its broad mouth margin is provided with barbels; the fin rays of its dorsal and pectoral fins fuse into a stout, serrate, erectile spine. In North American forms armouring derm plates are developed only on the head roof (Fig. 177). Closely akin to these are the Asiatic genera, and the single European species, *Silurus glanis*, the gigantic *Wels* of the Danube. The Nile is of interest if only for its forms of catfish to parallel the shapes and structures of the recent Teleosts.

In South America the catfish is a regnant type, and is
remarkable for the variety as well as for the number and
size of its forms.   Many, completely armoured (Fig. 178),
are strongly suggestive of Ganoids.   Their armouring is

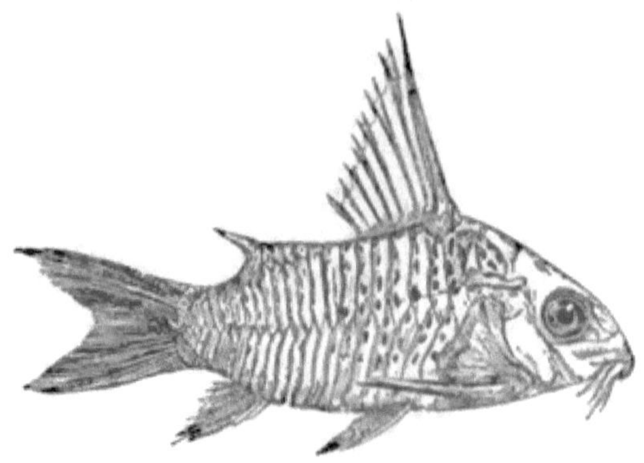

Fig. 178. — South American Siluroid, *Callichthys armatus.*   × 1.   (After
GÜNTHER.)   Upper Amazon.

metameral and archaic, their sensory canals primitive in
structure and arrangement.

Mormyrus, like the catfish, appears to have long been
divergent from the main stem of the Teleosts.   Its species

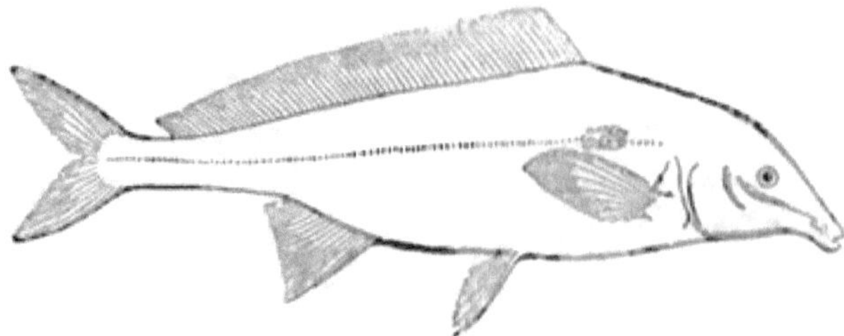

Fig. 179. — *Mormyrus oxyrhynchus.*   × ⅙.   (After GÜNTHER.)   Nile.

are restricted to the Nile, one — the long-nosed *M. oxyrhynchus* (Fig. 179) — figuring prominently in Egyptian myth.
In many of its structures it is archaic, as in axial skeleton,
fins, dermal characters, sensory canals ; in others, *e.g.* hear-

ing organ, it is most highly specialized. Its group is an interesting one, and has been but little studied.

The *Eel* (Fig. 180) might well be taken as one of the

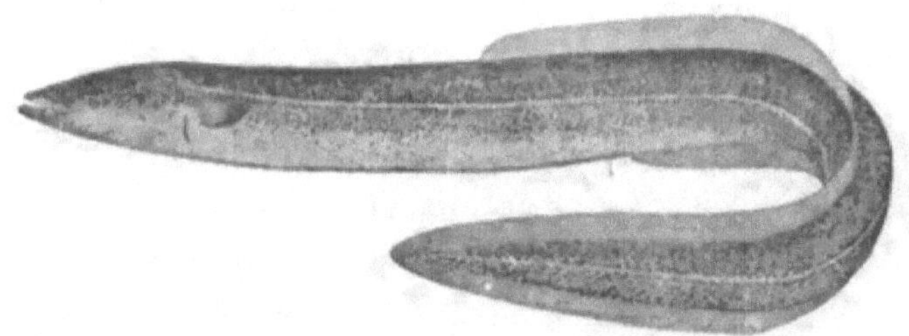

Fig. 180. — The eel, *Anguilla vulgaris*, Turton. × ¼. (After GOODE in U. S. F. C.) Europe, South Asia, North Africa, North America.

fish forms evolved by special environment. Living in soft river bottoms, a serpent-like movement in progression has gradually been acquired; its form has, therefore, become elongated and rounded, and the internal structures correspondingly modified. Fin structures have accordingly been

Fig. 181. — The perch, *Perca americana* (= *fluviatilis?*), Schrank. × ¼. (After GOODE in U. S. F. C.)

metamorphosed, ventral fins lost, tail degenerated, and a continuous dorsal and ventral secondarily evolved; scales have become reduced in size, supplanted by mucous layers.

Similarity in eel-like form, *e.g.* as of *Muræna*, is not in itself indicative of direct kinship. (*Apodes.*)

The *Perch* (Fig. 181) has long been taken as a representative Teleost. Perfect in its "lines," its compact, wedge-like shape cleaves the water by vigorous thrusts of a strong broad caudal; its fins are stout, supported by spinous rays; its dermal armouring light, smooth, and flexible; its colour is brilliant under its transparent scales. So adapted is it to its environment that its organ of static equilibrium, the air-bladder, has lost its valvular connection with the gullet. Of existing fishes about one-half are essentially percoid. (*Acanthopterygii.*)

Fig. 182.— The codfish, *Gadus morrhua*, L. × ¼. (After GOODE in U. S. F. C.) North Atlantic.

The *Cod* (Fig. 182) is scarcely less important as a representative Teleost. Its structural differences may perhaps represent the result of a competition less active than that of the perch in the struggle for survival. Heavy in body, its sluggish form has become blunted and rounded; its fins are depressed, their rays soft and yielding; its scales are reduced in size, colours less vivid; its swim-bladder loses its connection with the gullet. As many, perhaps, as one quarter of the existing genera of fishes may be assigned to this type. (*Anacanthini.*)

The *Flounder* (Fig. 183) should be mentioned as a singu-

lar instance of environmental evolution, its flattened body
adapting itself both in shape and colour to its bottom
living. Its entire side, — not the ventral region, as in the
rays, — is flattened to the bottom. The unpaired fins now
become of especial value ; they increase in size, and their
undulatory movements enable the fish to swim rapidly yet
retain its one-sided position ; ventral fins become useless,
and degenerate. The further adaptations of the flat fish
include its pigmentation only on the upper or light-exposed
side, and the rotation of the eye from the blind to the upper

Fig. 183. — The winter flounder, *Pseudopleuronectes americanus* (Walb.), Gill.
× ½. (After GOODE in U. S. F. C.) North Atlantic.

side, — in this giving one of the most remarkable cases of
adaptation known among vertebrates. (*Heterosomata.*)

The *Porcupine-fish* (Fig. 184) may be referred to as
another singular result of environmental evolution. Its
globular and inflatable form bespeaks slowness of motion
and helplessness if exposed to changes of temperature
or current. Its fins are reduced and feeble, suited, however, to its tranquil habitat ; its fused jaws, parrot-like,
show in how special a way its food is best secured. It
has evolved a protective casing of enormous needle-like
scales, whose shape parallels that of the derm denticles

of the shark.　As a somewhat transition form to the more usual conditions of the Teleost, the *Rabbit-fish* has been figured (Fig. 184 *A*).　(*Plectognathi.*)

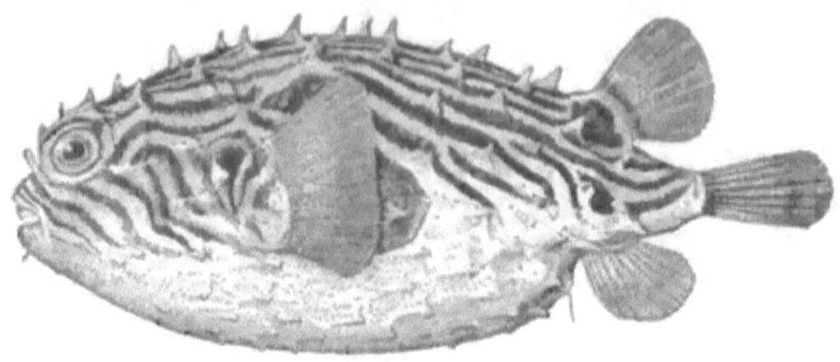

Fig. 184. — The porcupine-fish, *Chilomycterus geometricus* (Schn.), Kaup.　× ⅓. (After GOODE in U. S. F. C. report.)　Warmer Atlantic.

Fig. 184 A. — The rabbit-fish, *Lagocephalus lævigatus* (L.), Gill.　× ⅓.　(After GOODE in U. S. F. C.)　Northeast Atlantic.

A final, perhaps the most bizarre, instance of adaptation among Teleosts is that of the *Sea-horse* (Fig. 185). In spite of its many structural oddities, its genetic kinship with the Sticklebacks (Hemibranchiates) cannot be doubted.　Yet to have attained its present form its evolution must have been carried along a widely divergent path. It may, in the first place, have fused the lines of its metameral scales, dividing off the surface of its elongate body in sharp-edged rectangles, whose corners became produced as spines. At this stage of evolution its appearance might well be represented by (Fig. 185 *A*) the kindred *Pipe-fish*. To secure more perfect anchorage in its algous feeding-ground, its body terminal must now have discarded its fin membranes and become *prehensile,* — probably the most remarkable adaptation in the entire class of fishes, since it causes metameral organs to change the plane in which they function from a horizontal to a vertical one. As a probable development of prehensilism, three changes may next have been wrought : the flexure of the neck region, the thickening of the trunk, and the metamorphosis of the fins. The first change may have been brought about by the normal position of the fish's axis becoming, as is well known, vertical ; the head then assumes its normal horizontal plane and thus parallels mildly the cranial flexure of higher animals. The enlargement of the trunk region is evidently of static value. The alteration of the position, size, and degree of move-

Fig. 185. — The sea-horse, *Hippocampus heptagonus*, Raf. × ¼. (After GOODE in U. S. F. C.) East coast of North America.

ment of the pectoral fins, the loss of the ventrals and the changed function, now one of propulsion, of the dorsal, appear clearly the result of the altered plane of the fish's motion. Further structural changes might with interest

N

be followed, as in characters of viscera, gills, and endo-
skeleton.   In its life habits mimicry is strongly evinced;

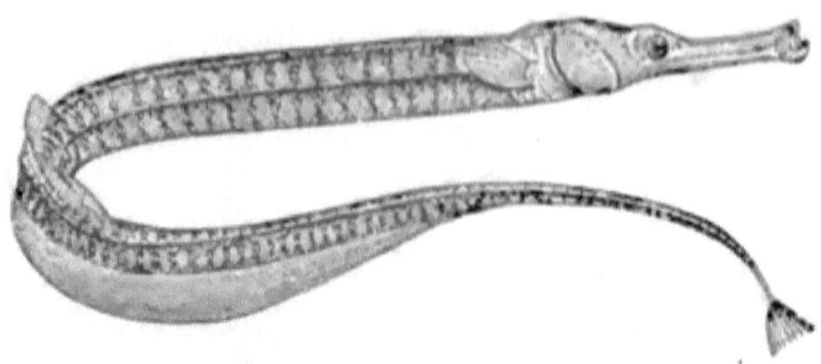

Fig. 185 A. — The pipe-fish, *Syngnathus acus* ♂, L., showing abdominal pouch.
× 1.   (After GÜNTHER.)   Coasts of Europe and Africa.

the well-known genus *Phyllopteryx*, whose entire body
surface develops pigmented appendages, is with difficulty
to be distinguished from a rough-shaped seaweed.   (*Lophobranchii.*)

VIII

THE DEVELOPMENT OF FISHES

THE groups of fishes have hitherto been contrasted
in the structures of their living and fossil forms. They
should next be reviewed in the light of their mode of
development; for the developmental stages of the Shark,
Lung-fish, or Teleostome might be expected, according
to time-honoured belief, to furnish important evidence
as to their descent and interrelationships. The younger
stages of the various forms of fishes should thus suggest
their ancestral characters: the developing Teleost should
approach the Ganoid; the Lung-fish and the Ganoid
should resemble their supposed elasmobranchian ancestor.

But the embryology of fishes is in this regard very
inconclusive, if at present in any important way suggestive. The majority of the forms, including some of the
most important, are developmentally unknown; yet sufficient is known of the representative members of the
groups to show the most perplexing characters. On the
one hand, the developmental processes of forms which are
regarded by the morphologist as closely akin seem often
widely distinct; and, on the other hand, the fishes which
should, *a priori*, exhibit an archaic mode of development
actually present complex processes of early growth which
can only be interpreted as highly specialized. In fact,
there are far greater differences in the developmental plans

179

of the closely related Ganoid and Teleost, than in those of
a Reptile and a Bird; and even among the members of the
single group, Teleosts, there are more striking embryolog-
ical differences than those between Reptiles and Mammals.
Adaptive characters have entered so largely into the plan
of the development of fishes that they obscure many of
the features which might otherwise be made of value for
comparison. And until the controversies regarding some
of the most fundamental principles in embryology — *e.g.*
the importance of the loss or gain of food yolk — shall be
decided, it seems impracticable to use the plan of development as in any strict sense a guide in phylogeny.

It is, accordingly, rather with the view of contrasting the groups of fishes, whose external features have
hitherto been compared, that the present chapter seems
of especial importance. They may briefly be reviewed in
their (A) spawning·habits, (B) the mode of fertilization
of their eggs, (C) their embryonic, and (D) larval development.

A. EGGS AND BREEDING HABITS

The eggs of typical fishes in Figs. 186–199, illustrate
how wide a range occurs in their shapes and sizes. All
are of about actual size, except Figs. 189–191, which have
been reduced about two-thirds. From the figures the
character of the egg membranes may also be contrasted.

Among Cyclostomes, which are usually looked upon
as of close genetic kinship, there appears a striking difference in the characters of the eggs. Those of *Bdellostoma* and *Myxine* (Figs. 186, 187) are large and bluntly
spindle-shaped, encased in a horn-like capsule; those, on
the other hand, of *Petromyzon* are minute, spherical, and
enclosed in delicate and jelly-like membranes (Fig. 188).

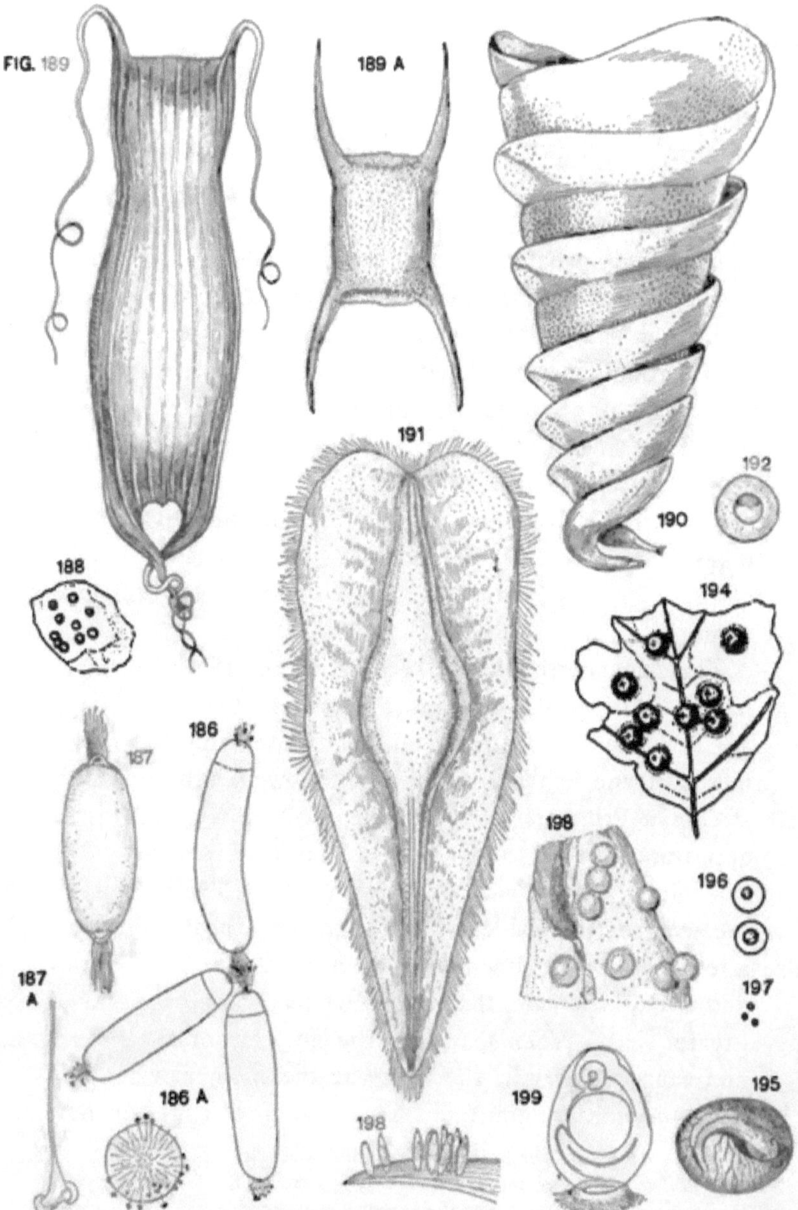

Figs. 186–199.—Eggs and egg cases of fishes. All of about actual size except 189–91; these have been reduced about two-thirds. 186. *Bdellostoma*, germ disc (?) at upper pole and in 186 A terminal hook processes and micropyle. (After AYERS.) 187. *Myxine*. (After STEENSTRUP.) 187 A. Terminal process. 188. *Petromyzon marinus*. 189. Shark, *Scyllium*. (After GÜNTHER.) 189 A. Skate, *Raja*. 190. Port Jackson shark, *Cestracion*. (After GÜNTHER.) 191. Chimæroid, *Callorhynchus*. (After GÜNTHER.) 192. Lung-fish, *Ceratodus*. (After SEMON.) 193. Ganoid, *Lepidosteus*. 194. Ganoid, *Acipenser*. 195. Siluroid, *Arius*, showing larva. (After GÜNTHER.) 196. Teleosts: **sea-bass**, *Serranus*, and 197. shad, *Alosa*. 198. Blenny, *Blennius*, showing attached **egg capsules**. 199. Enlarged Blennius (after GUITEL), showing mode of attachment of capsule.

The eggs of Myxinoids are probably deposited at a single time ; at first extruded by pressure of the body wall ; then drawn out string-like, one egg following another, attached by hooked and thread-like processes (Figs. 186 *A*, 187 *A*). Little is known, however, of the actual breeding habits of Myxinoids, either as to locality, mode, or season ; individuals of Myxine and Bdellostoma with ripe spawn have never been taken even in the most favourable regions. It is supposed that their spawning does not occur in the immediate neighbourhood of the shore, since detached eggs have been dredged in the deeper water. Their breeding time is probably in the early spring, although possibly intermittent spawning takes place. In Myxine, according to Putnam,[*] the bulk of the eggs may be deposited as late as the beginning of winter.

The spawning habits of Petromyzon, on the other hand, have been especially favourable for observation. The eggs are deposited in shallow and clear water and the movements of the fish may readily be followed. In the small stream at Princeton,[†] for example, the lampreys make their appearance about the middle of May and remain on the spawning grounds two or three weeks. Their "nests" are seen scattered thickly on the gravelly shoals, often but a few feet apart. Each will be occupied by several males and a single female, the latter conspicuous on account of greater size. When spawning, the lampreys press together and cause a flurry in the water at the moment when the eggs and milt are emitted. This portion of eggs will now

---

[*] As observed at Grand Menan. Pro. Bost. Soc. Nat. Hist. Feb. '74.

[†] Professor McClure and Dr. O. S. Strong have here repeatedly observed the spawning lampreys; it is to their account that the writer is here indebted. Compare, also, the excellent account given recently by Professor Gage. *Ref.* p. 234.

be covered with a thin layer **of sand or gravel, — the** spawners always returning to the same nest, — **and a sec-** ond, third, and more tiers of eggs will **be added. When the** eggs have finally been deposited, **the nest is fortified** by a dome-like mass of pebbles **and stones,** which the lam- preys carefully drag **to the spot.** The nest is thus marked out as well as protected, and is said to be made of **partial use during** the following **season.** The hatching **of the** eggs takes place within about a fortnight.

The eggs **which Sharks and Rays** deposit **are usually** enclosed in **a stout, horn-like capsule; this is in** general **of** oblong or rectangular outline, its **surface smooth or ridged;** the **case of** the **egg of** *Scyllium* (Fig. **189), shows** thread- like terminal processes, while these in **the ray (Fig.** 189 *A*) are stout and spine-like. **A great variation may exist in** the size of the egg **and in the character of its** envelopes among the different **groups of Elasmobranchs. The egg** of the Port Jackson **shark,** *Cestracion* (Fig. 190), **is of enor- mous** size and possesses an extremely thick, spiral-rimmed, **pear-shaped** capsule; that of **the Greenland shark,** *Læmar- gus,* **is said to be** spherical and relatively small, **and to be** deposited unprotected **by** capsule.

The breeding habits **of Elasmobranchs are but imper-** fectly known. With the **exception, perhaps, of Læmargus,** the sexes copulate.* The clasping appendages **of the male** are inserted either singly or together into the **cloaca and** oviduct of the female, and the eggs appear **to be** fertilized in the uppermost **portion of the oviduct. The egg then** becomes surrounded **by a glairy albuminous envelope, and** thereafter **by the secretion of the oviducal** gland, which in the **lower** oviduct **hardens into the horny capsule. The**

---

* The copulation of sharks has **been but rarely observed** (*e.g.* by Bolau in Hamburg; cf. *Ref.* **on p. 241).**

majority of sharks and rays are viviparous; the eggs are retained in the lowermost portion of the oviduct (uterus) and the embryo establishes a "placental" circulation, the vascular yolk sac becoming adherent to the walls of the uterus.   Other sharks deposit their eggs, and their mode of oviposition has been observed.   The egg (Fig. 189), when slightly protruded from the cloaca, is rubbed against brush-like objects, and when its terminal processes become finally entangled, the egg is withdrawn.   The processes of the egg case which leave the body last, the longer ones, are often greatly straightened out when the egg is deposited; subsequently their elastic character causes them to curl tightly, and often to secure a firm attachment to neighbouring objects.   The eggs of oviparous skates (Fig. 189 *A*) are said to be deposited on sand flats near the mark of low water.   Mr. Vinal N. Edwards of Wood's Holl, Massachusetts, believes that they are implanted vertically in the sand, and, from the occurrence of "beds" of skate eggs, that the fishes are singularly local in their places of spawning.   Eggs of Elasmobranchs* are often many months in hatching; the young fish finally escapes through a slit at the end of the egg case.

Nothing is known definitely of the breeding habits of Chimæroids.   The mode of copulation of the sexes is doubtless similar to that of sharks.   Their clasping organs are highly specialized sperm ducts, and the hook-bearing organs at the anterior margin of the ventral fin, and on the forehead of the male, function in all probability in retaining the female.   The forehead spine could certainly prove of such service if the position of the fishes during mating was at all similar to that figured for Scyllium by

---

* In the case of *Scyllium* the eggs are deposited about six days after they have been fertilized; they then hatch in from 200 to 275 days.

Bolau.* The egg case of Callorhynchus (Fig. 191) is essentially shark-like; it is of spindle-shaped outline, and its broad, fringing margin gives it an almost seaweed-like appearance. The egg is believed to be deposited in deep water.

The spawning of but one of the three existing Lung-fishes has been recorded. Ceratodus, according to Semon, has a spawning season extending over several months; it deposits its eggs in shallow water, scattering them broad-cast. The female fish is attended by several males, and the emission of eggs and milt appears to be simultaneous. The egg (Fig. 192) lacks a horny capsule, but is amply protected by a thick, jelly-like hull. It hatches during the second week.

Eggs of Ganoids are shown in Figs. 193, 194. They are encased in a jelly-like envelope, especially viscid in the case of sturgeon. When deposited, they speedily adhere to whatever they touch, and often remain attached until the time of hatching. The spawning grounds are in shallow water; the fish occur in numbers during a few days of May and June, each female attended by several males: ova and milt are emitted simultaneously, at short intervals. The eggs develop rapidly, hatching in about a week.

The eggs of Teleosts present the utmost variety in number, form, membranes, and mode of deposition. In some forms (Embiotocids, Blenniids, Cyprinodonts) they may even develop within the ovarian tissue, establishing there a "placental" circulation. They have been fertilized within the fish, the anal fin spine of the male having in some cases been metamorphosed into a copulatory organ. The eggs of Siluroids (Fig. 195) are generally of large size,

* V. *Ref.* p. 241.

and somewhat adhesive; they are deposited in "nests," *i.e.* bowl-like depressions, and are attended by the male fish.*
Other adhesive eggs are those of carp, *Christiceps, Batrachus.* Eggs of Salmonids are deposited loosely in "nests" on a clean, gravelly bottom; their membranes are thick and parchment-like. On the other hand, the majority of pelagic fishes produce eggs which float (Figs. 196, 197); of these the membranes are extremely hygroscopic and transparent, and an oil globule, located in the yolk region of the egg, serves to diminish its specific gravity. The egg membranes of a number of Teleosts, *e.g.* Blennies (Fig. 199), appear essentially shark-like; a horn-like capsule is evolved, whose terminal processes afford it a firm attachment. Aberrant modes of oviposition are not lacking; the South American Siluroid, *Aspredo,* as is well known, carries its eggs attached to its ventral surface; the pipe-fishes and sea-horses, *Siphostoma, Solenostoma, Hippocampus,* have specialized a pouch-like fold of the abdomen and of the ventral fins, which serves to retain the eggs and larvæ. It is curious to note that this remarkable condition occurs only in the *male.*

The breeding habits of Teleosts are in general like those of Ganoids; their spawning season is usually during the spring and summer, but is seldom of very brief duration. The hatching of the eggs depends largely upon water temperature, and may vary from a few days to several months (Salmo).

B. THE FERTILIZATION PHENOMENA

The processes of the maturation and fertilization of the egg have as yet shown but minor differences in the

---

* In several genera they are carried about in the gill chamber of the male, thus ensuring aëration.

groups of fishes. In the forms which have thus far been studied* there have been few noteworthy variations from what appear the normal conditions of vertebrates. The sperm usually gains admission to the egg through a micropyle in the egg membranes which becomes formed immediately after the extrusion of the polar bodies. A sperm cell, invariably a single one, participates in the actual fertilization. This may occur directly by the formation of a single male pronucleus, as *e.g.* in Petromyzon, Teleosts; while in the sharks, on the other hand, Rückert describes a multiple fertilization (polyspermy), where many male pronuclei† are formed, the one nearest in position fusing subsequently with the female pronucleus. An intermediate condition seems to be retained in the sturgeon, where several (six to nine) micropyles have been noted, although but a single one occurs in the kindred Ganoid, Lepidosteus (Mark, *Ref.* p. 249).

C. THE EMBRYONIC DEVELOPMENT

When the egg of a fish is deposited, it contains but the elements of a single cell. Its size and its enveloping membranes may vary widely, but its constituents are constant, — cytoplasm and nucleus. The size of the egg in different fishes varies with the amount of food material, or yolk, stored away in its cytoplasm; the enormous egg of the shark differs from the minute egg of the lamprey strikingly in this regard. But even in the minute lamprey egg there is a certain amount of yolk material present.

In every egg there can usually be distinguished at sight

---

* Lamprey by Kupffer and Böhm, and **Calberla**; **Sharks by Rückert**; Teleostomes by **Hoffman**, Agassiz and Whitman, Kupffer, **Böhm**, and others.

† These appear later to undergo karyokinesis, and are thereafter to be regarded as supplemental merocytes (p. 195).

an upper and a lower zone : the latter rich orange in colour, caused by the settling of the heavier yolk material ; the former lighter in colour, containing the nucleus of the egg, and originating the growth processes.

The less the amount of yolk in the lower, or vegetative, region, the smaller is naturally the egg, and the more obscure becomes the limit of the upper zone, or germ, or animal pole, as it is indifferently called. In the yolk-filled egg of the shark, on the other hand, the upper zone becomes reduced to a mere "germ disc" on the surface of the egg (Fig. 216, *GD*). If but little yolk is present, the early growth processes, *i.e.* the splitting of the germ cell, or egg, into many cells, or blastomeres, to give rise to the embryo, affect the entire egg. If, however, much yolk is present, the cells at first multiply only at the animal pole, and the yolk-filled region, remaining unsegmented, furnishes the nutriment for the cell growth above.

In the present outline of the development of fishes, the following types are reviewed : —

I. Petromyzon ; II. Shark ; III. Lung-fish ; IV. Ganoid ; V. Teleost.

I. *The Development of Petromyzon*

The egg of Petromyzon is of small size (Fig. 188), and is poorly provided with yolk material ; in surface view one can only distinguish the germinal from the yolk region by its slightly lighter colour. In the side view of the egg of Fig. 200, the beginning of the first cleavage plane is seen ; a vertical plane, passing through the egg, completes the stage of the two blastomeres of Fig. 201. The nuclei were at first close to the upper, or animal, pole, but they shortly take their position somewhat above the plane of the egg's equator. A second cleavage plane is again vertical, ap-

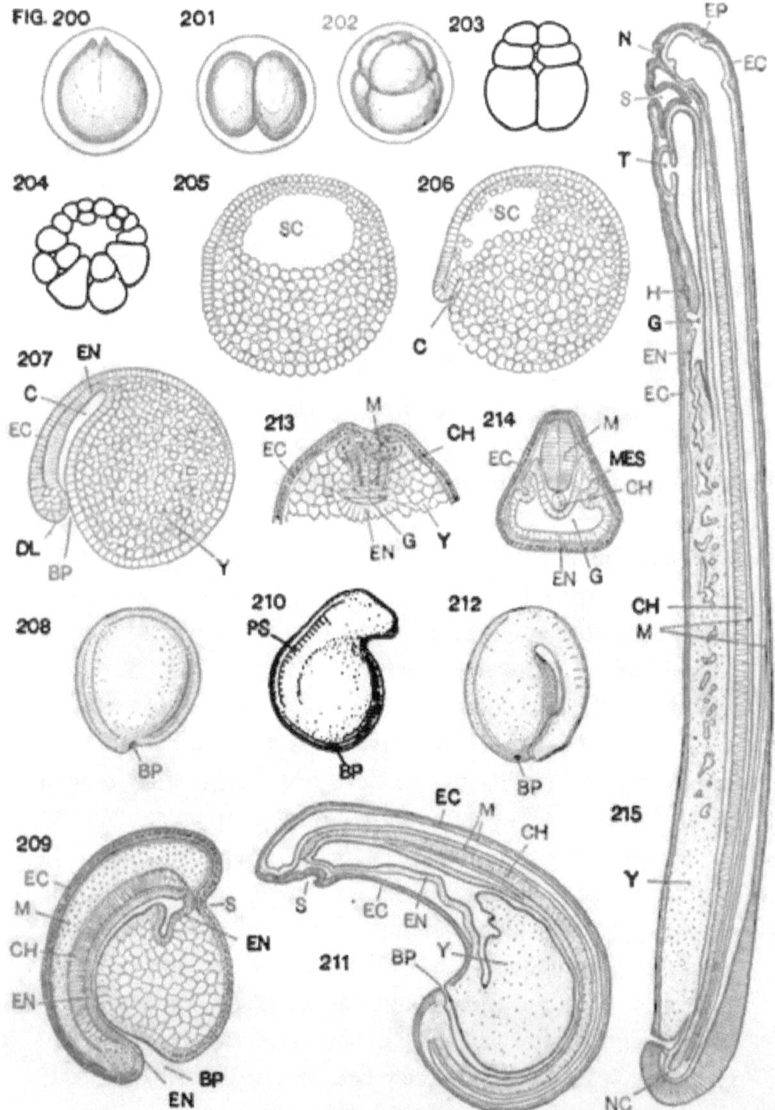

Figs. 200–215.—Development of lamprey, *Petromyzon planeri*. Figs. 200–204, 208–212 × 18, others × about 30. 200, 201. First cleavage, beginning and concluded. 202. Third cleavage. 203. Fourth cleavage, in section, showing beginning of segmentation cavity. 204, 205. Early and late blastulæ, in section. 206, 207. Early and late gastrulæ, in section. 208, 210, 212. Early embryos showing growth of head end. 209, 211. Sagittal sections of early embryos showing differentiation of organs. 213, 214. Transverse sections of early embryos. 215. Sagittal section of newly hatched larva, *Ammocœtes*. (Figs. 211, 215, after GOETTE, others after v. KUPFFER.)

BP. Blastopore. *C.* Cœlenteron. *CH.* Notochord. *DL.* Dorsal lip of blastopore. *EC.* Ectoderm. *EN.* Entoderm. *EP.* Epiphysis. *G.* Gut. *H.* Heart. *M.* Central nervous system. *MES.* Mesoblast. *N.* Nasal pit. *NC.* Neurenteric region. *S.* Mouth pit, stomodæum. *SC.* Segmentation cavity. *T.* Thyroid gland. *Y.* Yolk and yolk cells.

proximately at right angles to the first ; the third, which shortly appears, is horizontal (Fig. 202), giving rise to the stage of eight blastomeres ; this plane, passing slightly above the equator, causes the upper blastomeres to be slightly smaller in size than those of the lower hemisphere. The amount of yolk in the egg, it is accordingly inferred, although not sufficient to prevent the passage of cleavage planes, is enough, nevertheless, to retard the nuclear cleavages in the region of the lower, or vegetative, pole. In Fig. 203, showing a vertical section of the following stage, another horizontal cleavage has been established in the upper part of the egg ; the segmentation cavity is seen in the centre of the figure arising as the central space between the blastomeres. This is seen to have become greatly enlarged in Fig. 204, a slightly later stage where in vertical section is seen a greatly increased number of blastomeres. Repeated cleavage of all blastomeres now continues regularly, and results in the production of a *blastula*, a smooth-surfaced cell mass containing the segmentation cavity, SC (in section, Fig. 205) ; this is seen to be located in the region of the animal pole. In the next developmental stage, *gastrula*, seen in section in Fig. 206, the primitive digestive tract, *cœlenteron*, C, is appearing ; it arises as an indentation of the side of the blastula. The cœlenteron, soon greatly increasing in depth, reduces in size and finally obliterates the segmentation cavity, taking the position, C, shown in section in Fig. 207. Here the segmentation cavity has practically disappeared ; the surface opening of the cœlenteron is the *blastopore*, BP ; the cell layer of the gastrula's surface is the *ectoderm*, EC ; the cell layer lining the cœlenteron is the *entoderm*, EN : the cœlenteron, it will be seen, is closely apposed to the ectoderm at the left of the figure, — the future dorsal region of the embryo; on this side the margin of the blastopore is known as the dorsal lip, *DL*, while to the right the ventral lip is seen greatly enlarged by the yolk-bearing cells, *Y.* A somewhat later stage (Fig. 208) shows the blastopore as a narrowly constricted opening, *BP*, whose dorsal lip is slightly raised at its left-hand margin. The head of the embryo is to arise near the opposite pole (as in Fig. 210), and is thence to elongate into neck and trunk (Fig. 212). A sagittal section of a stage, slightly older than Fig. 208, shows admirably the structures of the embryo that have thus far been differentiated (Fig. 209). Contrasting with Fig. 207, it will thus be seen that the cœlenteron, arising at *BP*, has become greatly elongated; at its blind end its lining membrane, entoderm, *EN*, is in contact with an indented portion of the ectoderm, at *S*, where later the opening of the mouth will be established; and that ventrally the cœlenteron has given off a pouch which passes into the yolk, and will later be differentiated as the liver. That the entire dorsal wall of the cœlenteron has become thickened, constitutes the main difference between the sections of Figs. 207 and 209; there have, in other words, arisen between the entoderm and ectoderm of Fig. 207 the central nervous system, or medullary cord, *M*, and the notochord, *CH*. The origin of these structures may best be traced in the cross-section of a slightly earlier stage (Fig. 213); the cœlenteron, or gut, is at *G*, the ectoderm at *EC*, the yolk cells intervening at *Y*; and the notochord and medullary cord, *CH*, and *M*, in the sagittal region immediately between the gut and the ectoderm. In the medullary region the ectoderm cells are seen pressed together, growing downward and sidewise, forming altogether a compact cell cord *

* As in Teleosts, but unlike other vertebrates.

passing down the back of the embryo ; the notochord is arising from the differentiating cells of the roof of the gut. In
the cross-section shown in Fig. 214, the subsequent conditions of these structures may be seen ; the medullary
nerve cord, *M*, is now in section elliptical, separated dorsally from the ectoderm, and its cellular elements are of
more uniform size, arranged with bilateral symmetry, its
central lumen having not as yet appeared ; the notochord,
now constricted off from the wall of the gut, takes upon
it its characteristic form and structure. It is, however,
in the differentiation of the walls of the gut that this
section is of especial interest ; the gut is seen to have
greatly enlarged, and at the expense of the yolk material ;
its lining membrane, entoderm, *EN*, is now directly apposed to the outer germ layer, ectoderm, *EC.* The middle
germ layer, *mesoderm*, *MES*,—out of which cartilage,
muscular and connective tissue, are formed,—is now seen
taking its origin as paired evaginations of the dorsal wall
of the gut. The mesoderm shortly loses its connection
with the entoderm, and by the rapid increase of its cellular
elements rapidly invests the remaining embryonic structures ; its segmental character may be seen in the surface
view shown in Fig. 210, its dorsal portions appearing as
the primitive segments.

Later developmental stages are shown in the sagittal
sections, Figs. 211, 212. These may best be compared
with Fig. 209. In Fig. 211 the head end of the body has
greatly elongated, and with it the gut cavity has dilated ;
entoderm is now composed of very minute cells, whose
nuclei are suggested by dots ; the yolk has become more
definitely restricted to the region of the hinder gut ; the
blastopore is still seen ; at its lips the germ layers are
alone fused.

II. *The Development of the Shark*

On the side of embryology a shark presents many points of striking contrast to the lamprey; yet it may in many regards be looked upon as archaic in its developmental characters. Its contrasting structures (together with those of lung-fish, Ganoid, and Teleost) may best be reviewed in the table, p. 280.

The egg of the shark is of large size, richly provided with yolk material. When removed from its membranes, it is seen to be of a bright orange colour; its form is elongated, and the weight of its pasty substance causes it to assume a flattened ovoid (Fig. 216). At the upper pole of the egg is a small, light-coloured spot, the germ disc, *GD*, which figures prominently in the early stages of development. It would represent the lamprey's entire egg, if one could imagine a point of the lower pole of the latter hugely dilated with yolk. It is in the region of this germ disc alone that every process of development as far as gastrulation occurs.

The segmentation of the germ disc is shown in Figs. 217–220. In the first of these (Fig. 217) the germ is seen to be sharply marked off from the surrounding yolk by a circular band; two cleavages have traversed it in the form of narrow grooves separating the blastomeres. In Fig. 218 the fifth cleavage has been completed; the furrows dividing irregularly the surface of the germ disc fade away at its periphery. Fig. 219 represents a vertical section of the germ disc at this stage; the upper, finely dotted layer, thinning away at either side, is the germ disc; the coarsely granular material below is the yolk; the depth of the cleavage furrows is seen, and it will be noted that up to this stage of development there have been no horizontal

o

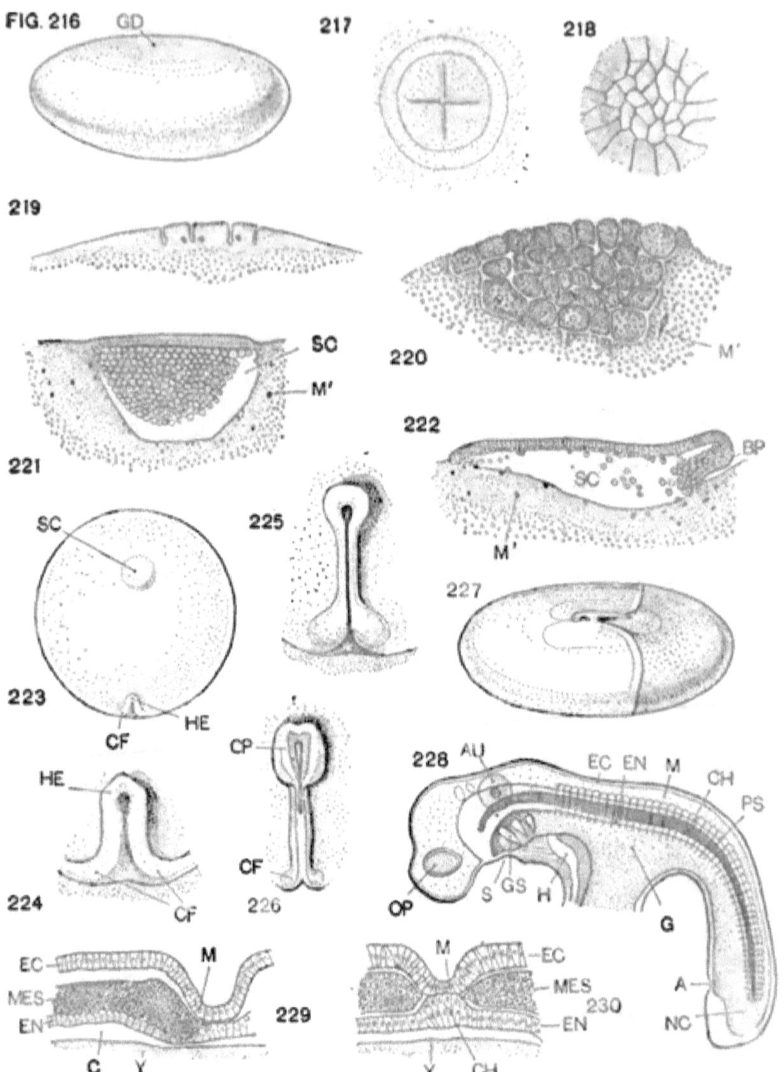

Figs. 216-230. — Development of shark, *Scyllium* (mainly). (All but 216 after BAL-
FOUR.) 216. Egg freed from case showing germ disc *GD*. 217. Germ disc at second
cleavage. 218. Germ disc at fifth (?) cleavage. 219. Vertical section of similar stage.
220. Vertical section of slightly older germ disc. 221. Blastula. 222. Early gastrula.
223. Blastoderm showing early growth of embryo. 224-226. Slightly later stages of growth
of embryo. 227. Stage showing early embryo and mode in which the blastoderm surrounds yolk. 228. Early embryo viewed as a transparent object. 229, 230. Transverse
sections of early embryo.

A. Anal invagination. *AU.* Auditory vesicle. *BP.* Dorsal lip of blastopore. *C.*
Cœlenteron. *CF.* Tail folds. *CH.* Notochord. *CP.* Cephalic plate. *EC.* Ectoderm.
E.N. Entoderm. *G.* Gut. *GD.* Germ disc. *GS.* Gill slits. *H.* Heart. *HE.* Head
eminence. *M.* Central nervous system. *M'.* Yolk nuclei, merocytes. *MES.* Mesoblast.
NC. Neurenteric canal. *OP.* Optic vesicle. *PS.* Primitive segments. *S.* Mouth pit,
stomodæum. *SC.* Segmentation cavity.

cleavages. A stage in which early horizontal cleavages are represented is shown in Fig. 220. This may well be compared with the last figure; the germ disc, while not increasing in diameter, is now seen to have multiplied its blastomeres by horizontal cleavages; it is converted into a plug-shaped mass of cells, sunken into the yolk material. At M' are cell nuclei, which have found their way into the adjacent yolk, and which there acquire a developmental importance. They become the so-called merocytes, or yolk nuclei.

The section of the germ shown in Fig. 221 represents a subsequent stage of development; the blastomeres, by continued subdivision, have become greatly reduced in size, and are clearly to be distinguished from the smooth-surfaced, yolk-like material lying beneath. Merocytes, M', are apparent in the superficial layer of the yolk; they are supposed to serve a twofold function, — on the one hand, to elaborate the yolk material and fit it for the embryo's use; on the other, to supply the cells which are being continually added to the germ's margin. In the figure a large cavity is shown to exist between the yolk and the mass of blastomeres. This cavity has been identified as the segmentation cavity, SC, and the developmental stage as the blastula; it is as though the lower hemisphere of the lamprey's blastula (Fig. 205) had become enormously enlarged, and all traces of the cells in the floor of its segmentation cavity lost, except in the layer of the metamorphosed cells, the merocytes.

In the next growth process the extent of the germ area becomes greatly increased; the thick blastula is now thinned out into a surface layer of regular cells, an enlarging disc-like *blastoderm*, which will eventually grow around and enclose the entire egg. The blastoderm of

Fig. 223 is a pale-coloured circular membrane of about a half inch in diameter lying on the surface of the egg. Sectioned at an earlier stage (Fig. 222) the blastoderm is seen to present the following contrast to the blastula of Fig. 221 : the floor of the segmentation cavity has flattened, and a sharp rim forms the outline of the blastoderm ; at one side this rim is seen to protrude over the yolk mass, leaving a narrow, fissure-like cavity between. This stage is identified as the gastrula ; the fissure-like cavity, the cœlenteron ; its marginal blastoderm, the dorsal lip of the blastopore ; its ventral lip, the entire yolk mass.

The growth of the embryo's form takes its origin at the blastopore's dorsal lip. In Fig. 223 the rim of the blastoderm is seen indented near the point *CF*, and its thickening at this region becomes more and more marked in subsequent stages ; on the other hand, the anterior portion of the blastoderm, growing continually on all sides, becomes excessively thin, flattening itself tightly to the yolk, and reducing the segmentation cavity to the small area indicated at *SC*. The growth of the embryo in the mid-region of the blastopore's dorsal lip may next be followed in the stages, Figs. 224, 225, 226. The indentation of the rim may thus be seen to assume a creeselike thickening, thrusting forward its blunt end, the head eminence, *HE*, over the blastoderm ; at the points *CF*, the tail eminences, the rim of the blastoderm is thick, protruding, appearing to be pressing together in the median line, and causing the body of the embryo to be actually pushed into form and thrust above the level of the blastoderm. In Fig. 225 the sides of the embryo are separated dorsally by a deep groove, the medullary furrow, the future canal of the central nervous system. In Fig. 226 this is seen at a more advanced stage ; its hinder portion has been roofed over by the coalesced sides, and the process of enclosing the groove is being continued anteriorly, although the head end of the embryo is now flattened out as the prominent cephalic plate.

In the stage figured in 227, the form of the embryo has been acquired: the head in the manner already outlined, the tail by the coalescence and subsequent outgrowth of the tail folds, *CF*. The entire embryo now rises above the blastoderm, as this continues to enclose the yolk. In the figure the yolk has thus been more than half enclosed; its final appearance is seen in the oval space outlined by a dotted line behind the embryo.

The origin of the germ layers is not as readily traced as in the Cyclostome. Ectoderm is the most clearly marked; even in the blastula (Fig. 221) it has appeared as an outer single-celled stratum clearly differentiated from the underlying cells. Entoderm is only to be seen on the dorsal wall of the cœlenteron: the ventral entoderm (cf. Fig. 222) is merged with the yolk. Mesoderm takes its origin from the inner layer on either side of the median line, but it arises as a solid cell mass instead of as the pouch-like diverticula in Petromyzon. Cross-sections of an embryo represented by Fig. 224 have been figured in Figs. 228 and 229; the former is of the hinder region and illustrates the mode of growth of the mesoderm, *MES*; the latter across the head region, shows that in this region the mesoderm is separated from the inner layer. Both sections show the simple character of the medullary groove, and the latter section the mode of origin of the notochord, *CH*, *i.e.* as an axial thickening of the entoderm.

An embryo of about the stage of Fig. 227 is extremely delicate and may readily be viewed as a transparent object.

By this time (Fig. 230) it will be seen that its prominent organs have already been differentiated. There are thus: medullary canal, *M*, with optic, *OP*, and auditory, *AU*, vesicles; gut with gill slits, *GS*, neurenteric canal, *NC*, and suggestion of mouth, *S*, and anus, *A*; notochord, *CH*; segmented mesoderm (primitive segments), *PS*, and heart, *H*. The medullary groove was converted into a canal, as has been already suggested, by the overroofing and fusion of the summits of the medullary ridges; its anterior dilatation is the brain; the gut, *G*, communicates freely below with the yolk mass; it is a cavity, a portion of the cœlenteron that has been constricted off with the embryo; its openings, the mouth, anus, and gill slits, are secondary, acquired after there have been established in these regions fusions of entoderm and ectoderm; the neurenteric canal, *NC*, a communication between medullary tube and gut, is·a structure acquired in the stage of Fig. 226, where the hinder medullary groove was roofed over, allowing, in the region of the tail folds, a communication to exist between medullary canal and cœlenteron. The notochord has by this stage been completely separated from the entoderm; it already assumes a supporting function.

III. *The Development of Ceratodus*

The development of a Lung-fish has thus far been described (Semon) only from the outward appearance of the embryo. The egg of Ceratodus (Fig. 192) is seen without its covering membranes, enlarged, in Fig. 231. Its upper pole is distinguished by its fine covering of pigment. The first fine planes of cleavage are shown in Figs. 232–236; and from these it will be seen that the yolk material of the lower pole is not sufficient to prevent the egg's total seg-

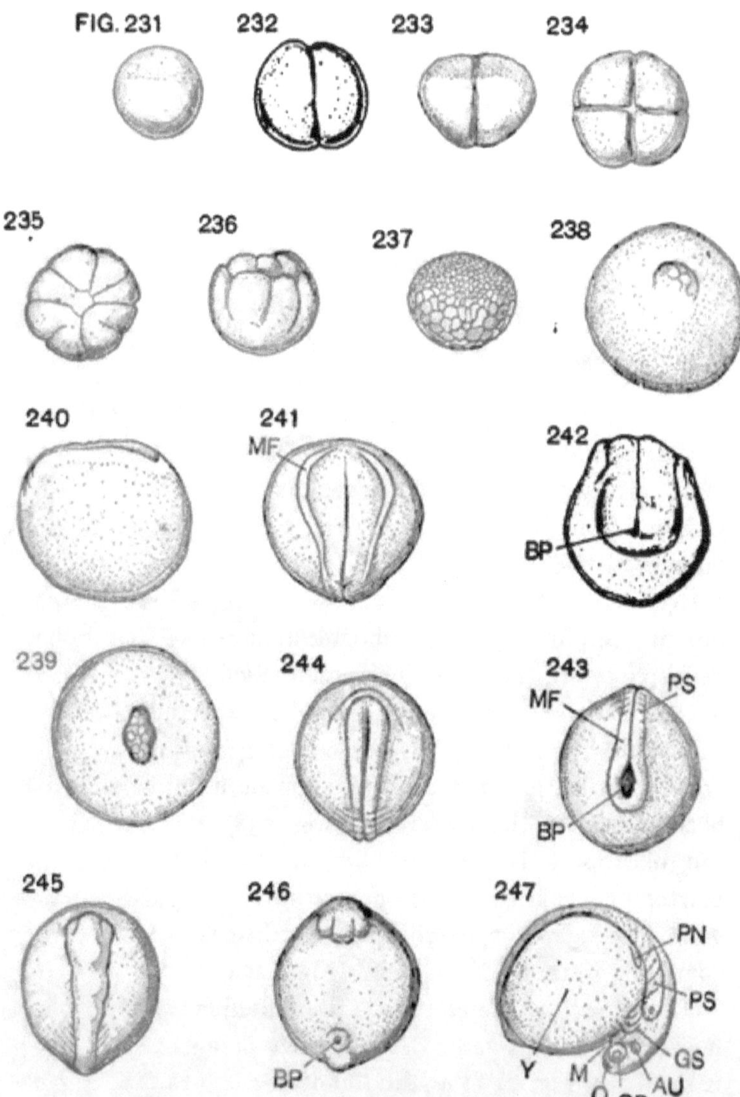

Figs. 231-247. — Development of lung-fish, *Ceratodus.* (After SEMON.) × 4-7.
231. Egg immediately before cleavage. 232, 233. First cleavage, seen from above and from the side. 234. Second cleavage, seen from above. 235, 236. Third cleavage, seen from above and from the side. 237. Blastula. 238, 239. Gastrulæ showing closure of blastopore. 240. Early embryo, seen from the side. 241. Early embryo showing medullary folds (head). 242. Tail region of same embryo. 243. Tail region of slightly later stage. 244. Head region of same embryo. 245-247. Later embryos. *AV.* Auditory vesicles. *BP.* Blastopore. *GS.* Gill slits. *M.* Mouth pit. *MF.* Medullary folds. *O.* Olfactory lobes. *OP.* Optic vesicles. *PN.* Primitive kidney, pronephros. *PS.* Primitive segments. *Y.* Yolk mass.

mentation. The first plane of cleavage is a vertical one, passing down the side of the egg (Fig. 233) as a shallow surface furrow, not appearing to entirely separate the substance of the blastomeres, although traversing completely the lower hemisphere (Fig. 232). A second vertical furrow at right angles to the first is seen from the upper pole in Fig. 234; it is essentially similar to that of Fig. 233. The third cleavage of Fig. 235 is again a vertical one (as in all other fishes, but unlike Petromyzon), approximately meridional; its furrows appear less clearly marked than of earlier cleavages, and seem somewhat irregular in occurrence. The fourth cleavage is horizontal above the plane of the equator. Judging from Semon's figure (Fig. 236), at this stage the furrows of the lower pole seem to have become fainter, if not entirely lost. A blastula showing complete segmentation is seen in Fig. 237; the blastomeres of the upper hemisphere are the more finely subdivided; the conditions of the segmentation cavity may be expected to prove similar to those of Fig. 205. Two stages of the gastrula are shown in Figs. 238 and 239, showing a full view of the blastopore. In the earlier one (Fig. 238) the dorsal lip of the blastopore is crescent-like; in the later (239) the blastopore acquires its oblong outline, through which the yolk material is apparent; its conditions may later be compared to those of a Ganoid (Figs. 254, 255).

The growth of the embryo is illustrated in the remaining figures (Figs. 240–248). A side view of an early embryo is shown in Fig. 240; at the top of the egg to the right is the head region, to the left the blastopore and tail. The surface view of the head region (Fig. 241), the medullary folds, *MF*, may be compared with those of Fig. 225, although they are low and widely separated; the axial seam is referred to by Semon as a demonstration of the theory of the embryo's **concrescence.** In the hinder **region of** the same **embryo** (Fig. **242**) **the** blastopore **is** still apparent, *BP*, reduced **to a narrow,** fissure-like **aperture;** **around it** is **the** tail mass, corresponding generally to *CF* **of** Fig. 226; and encircling **all is the hinder continuation** of the medullary folds.

The next change of **the embryo is strikingly amphibian-** like ; **the** medullary folds rise above **the egg's surface, and,** arching over, fuse their edges in the median dorsal **line.** In Fig. **243,** the tail region of **a** slightly older **embryo, this process is clearly** shown ; **the medullary folds,** *MF*, **are** seen closely apposed **in the median line ; hindward, how-** **ever, they are still** separate, **and through this opening the** blastopore, *BP*, may yet **be seen. At this stage primitive** segments **are shown at** *PS*; **in the brain region in Fig.** 244 the medullary folds are **still slightly separated (cf.** *CP*, Fig. 226).

Two **views of an** older embryo **are fig-** **ured** (Figs. 245 **and** 246), where the fish- like form may be rec- ognized. The medul- lary folds **have com-** **pletely** fused **in the** median line, **and the**

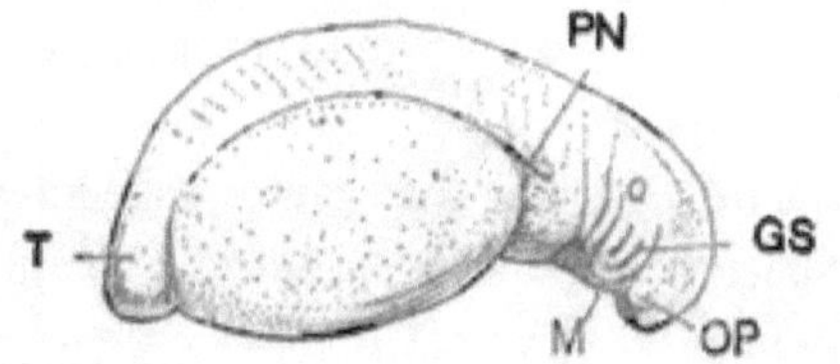

Fig. 248.—Embryo of Ceratodus, near the time of hatching.

GS. Gill slits. *M*. Mouth pit. *OP*. Optic vesi- cles. *PN*. Primitive kidney, pronephros. *T*. Tail eminence.

embryo **is** coming to acquire **a ridge-like prominence ;** optic vesicles and primitive **segments are apparent, and** at *BP* the blastopore appears to **persist as the anus. The** continued growth **of the embryo above the yolk mass,** *Y,* is apparent **in Fig. 247; the head end has, however,** grown the more rapidly, showing **gill** slits, *GS*, **auditory,** optic, **and nasal vesicles,** *AU*, *OP*, **and** *O*, **at a time when** the tail mass has hardly emerged from the surface. Pronephros has here appeared at *PN* (cf. with Fig. 247, Fig. 210). It is not until the stage of the late embryo of Fig. 248 that the hinder trunk region and tail come to be prominent. The embryo's axis elongates and becomes straighter; the yolk mass is now much reduced, acquiring a more and more oblong form, lying in front of the tail, *T*, in the region of the posterior gut (cf. Figs. 211 and 212). The head, and even the region of the pronephros, *PN*, are clearly separate from the yolk sac; the mouth, *M*, is coming to be formed.

IV. *The Development of Ganoids*

The development of Ganoids is next to be outlined. The eggs of the sturgeon and gar-pike are poorly provided with yolk. They have still, however, a greater amount than those of the lamprey or lung-fish, and in many regards of development suggest nearnesses to the Elasmobranchs.

The egg of the sturgeon shown in Fig. 249 shows clearly two distinct zones; the upper, blotched with pigment at the animal pole, is pale in colour; the lower, rich in yolk, is orange-coloured, well speckled with pigment. The early cleavages appear at first only in the upper pale-coloured area which corresponds apparently with the germ disc of the shark's egg. In Fig. 250 there have been two cleavages, vertical and at right angles to each other; these have sharply traversed the germ area, the earlier one being now produced slightly into the yolk region of the egg—only, however, as a slight surface furrow. The third cleavage (Fig. 251) presents a stage closely corresponding with that of Ceratodus of Fig. 235, its plane tending to pass parallel to the first cleavage: the germ disc

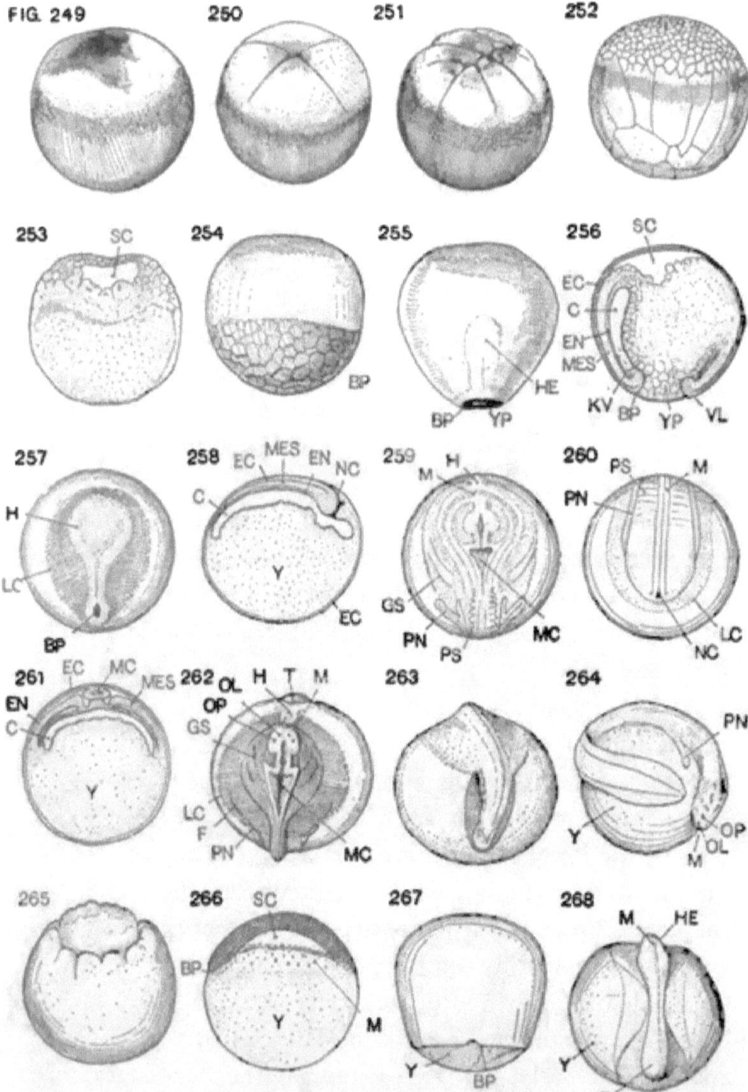

Figs. 249–268. — Development of Ganoids, *Acipenser* and (last four figures) *Lepidosteus*. × about 12. 249. Egg immediately before cleavage. 250. **Second cleavage**. 251. Third cleavage. 252. Blastula. 253. Vertical section of blastula. 254. **Early gastrula**. 255. Late gastrula. 256. Vertical section of late gastrula. 257. **Early embryo**. 258. Sagittal section of same stage. 259, 260. Head and **tail** regions of slightly later embryo. 261. Transverse body section of hinder body region of same stage. 262, 263. Head and **tail** regions of late embryo. **264.** Embryo immediately before hatching. 265. Lepidosteus' blastula. 266. Vertical **section** of early gastrula. 267. Late gastrula. 268. Embryo, showing mode **of** separation **from** yolk.

BP. Dorsal lip of blastopore. *C.* Cœlenteron. *EC.* Ectoderm. *EN.* Entoderm. *F.* Pectoral **fin.** *GS.* Gill slits. *H.* Heart. *HE.* Head eminence. *KV.* Kupffer's vesicle. *LC.* Marginal limit of cœlenteron. *M.* Mouth pit. *MC.* Medullary canal. *MES.* Mesoblast. *NC.* Neurenteric canal. *OL.* Olfactory pits. *OP.* Optic vesicles. *P.N.* Primitive kidney, pronephros. *PS.* Primitive **segments.** *SC.* Segmentation cavity. *T.* Tail eminence. *VL.* Ventral lip of blastopore. *Y.* Yolk, yolk mass. *YP.* Yolk plug.

is deeply cut by the furrows; the yolk area, however, only superficially; **the shallow furrow of the first** cleavage on **the yolk hemisphere** now passes **through the** lower pole; **the second** cleavage, **passing downward, has** made a shallow **groove** extending **half-way between the** rim of the **germ area and the lower pole of the egg. It is** the great **amount of·yolk in the lower hemisphere that retards** the **cleavage of the blastomeres. In** Fig. **252 the** entire **germ area has become subdivided into a** mass of small **cells, while the large, irregular blastomeres** of the yolk **hemisphere are separated only by superficial** furrows. **This stage, the blastula, is seen in section in** Fig. 253: **the yolk, unsegmented, occupies the lower** hemisphere; **the germ area contains a segmentation cavity,** *SC,* with **a roofing of small cells, and a floor of** irregular cells half **engulfed in a deep, underlying zone transitional** between **germ and yolk.**   ·

An early gastrula is seen in Fig. 254: the more rapid multiplication of the cells of the germ region has given **rise to a down-reaching cap of cells, whose** boundary is **here sharply marked off from the large and** imperfect yolk **cells of the lower hemisphere. At** *BP,* **the** rim of the cell **cap, or blastoderm, is sharply distinct from the yolk;** it is **the dorsal lip of the blastopore; the remaining** portion of **the rim is, generally speaking, the remainder** of the rim **of the blastopore; more accurately it is the** circumcrescence **margin of Hertwig. The late gastrula of** Fig. 255 **shows the greatly increased extent of the** blastoderm: its **margin is continually reducing the size of the** blastopore, *BP*; **on its dorsal lip at** *HE,* **the outline of** the embryo **is appearing. A sagittal section of this stage (Fig.** 256) **shows at** *BP* **the dorsal, and at** *VL* **the ventral,** lip of **the blastopore; at** *YP* **the yolk material appears at the** egg's surface as a plug-like mass; at *SC* is the segmentation cavity. **The** dorsal lip of the blastopore is seen to be far longer than the ventral **lip;** its **rim is the** more inflected, at *KV* occurring **a recessus which** the **writer compares to** the Kupffer's vesicle **of** Teleost development; the cavity, *C*, cœlenteron, **between the** wall of the blastopore and the yolk mass **is in this** region the largest. The germ layers in this stage, *EC*, *MES*, *EN*, are **seen to** be confluent **at** the blastopore's rim; at **the** termination of **the** cœlenteron, entoderm and mesoderm are merged; the ectoderm forms the **roof of** the segmentation cavity.

The form of **the embryo next becomes more definitely** established. **In Fig. 257 the blastopore, much reduced** in size, is seen at *BP*; its **thickened rim** is whitish in colour; the darkened **area, whose boundary** is *LC*, is **the** cœlenteron, seen faintly through the translucent margin of the blastopore; the embryo **is** the opaque **area** of **the** blastopore's dorsal lip, terminating anteriorly in the **dilated** tract, *H*, the head region. **In a** sagittal **section of a** slightly later stage **(Fig. 258), the** relations **of** germ layers, *EC*, *MES*, *EN*, cœlenteron, *C*, **and** yolk **mass,** *V*, **may be compared with those of** the section **(Fig. 256),** wherein the region *YP* corresponds to that **of** *NC*. **A** thin ectoderm will now be seen to have enclosed the entire egg; the segmentation cavity has disappeared; the rim of the blastopore, **becoming** continually constricted, causes the yolk material to **recede from** the surface, and leaves the blastopore disappearing, **as the** blunt diverticulum of *NC*. **The** neurenteric canal, *NC*, **is** the last communication **between the surface of** the egg and the cœlenteron; this has become established before the blastopore **closes in** the stage of Fig. **257 at** its dorsal lip;

the medullary furrow of the embryo has here been the deepest, and has been bridged over by a coalescence of its margins. At the anterior end of the embryo the inner, *EN*, and middle, *MES*, germ layers become greatly thinned, in the region where the heart is shortly to arise.

The next stage of development is represented in Figs. 259, 260, showing front and hinder regions of the same embryo. The curiously flattened mode of growth characteristic of the sturgeon is here very apparent; the embryo has surrounded over three-fourths of the egg's circumference, yet has not risen above its surface curvature; the head region is especially flattened; mouth, *M*, heart, *H*, gill slits, *GS*, brain, and optic vesicles are broadly spread out: the fourth ventricle at *MC*, the pronephros at *PN*, the primitive segments at *PS*. In the tail region the medullary folds appear at *M*, the pronephric duct at *PN*, the neurenteric canal at *NC*. A favourable section through the hinder body region of an early embryo is shown in Fig. 261; it illustrates the mode of origin of the following structures: the notochord as an axial thickening of entoderm, *EN*, immediately under *MC*; the medullary canal, as an infolding of (an under, or formative layer of) the ectoderm, its sides, folding over dorsally, coming to fuse in the median line; the mesoderm, *MES*, as in sharks, arising (partly) from the entoderm on either side of the notochord.

The later stage, shown in Figs. 262 and 263, may be contrasted with Figs. 259 and 260; the head region, though still greatly flattened out, is now rising above the surface; the trunk region is becoming prominent; the tail is budding out, and separating from the egg surface; sense organs are well outlined, and pectoral fins, *F*, elasmobran-

chian in character, are appearing. An embryo shortly
before hatching is next figured (Fig. 264); the head has
now entirely lost its flattened character; the mouth invagination occurs at *M*; the tail, much elongated, is
compressed laterally, and already presents the dermal
embryonic fin; the yolk sac is attached along the anterior body region, in a position more nearly that of the
shark than of the lung-fish.

Of the two Ganoids, sturgeon and gar-pike, the latter,
as the writer has pointed out, * has the more shark-like
developmental features. Its segmentation is incomplete,
since the yolk pole of the egg is at no time traversed even
by superficial furrows. The blastoderm, or cell cap, is
early apparent, and is clearly marked off by a furrow from
the irregular marginal blastomeres (Fig. 265). It resembles closely the segmented germ disc of an Elasmobranch,
and the irregular marginal blastomeres may be compared
to merocytes. The section of a late blastula of Fig. 266
does not differ widely from that of the shark of Fig. 221;
a segmentation cavity is present, whose floor is smooth,
and contains a well-marked zone of merocytes, *M*; the
smaller quantity and firmer consistency, perhaps, of the
yolk do not, on the other hand, permit the blastula to
occupy the sunken position of that of the shark. In the
gastrula of the gar, further, a well-marked notch appears
at the dorsal lip (as in this stage, Fig. 223, of the shark),
representing the primitive blastopore. And, finally, the
form of the embryo rises boldly from the surface, and
early presents the well-marked head and tail eminences,
HE and *T*, of Fig. 268, comparable with Figs. 225 and
227.

* *Am. J. Morph.*, Vol. XI, No. I.

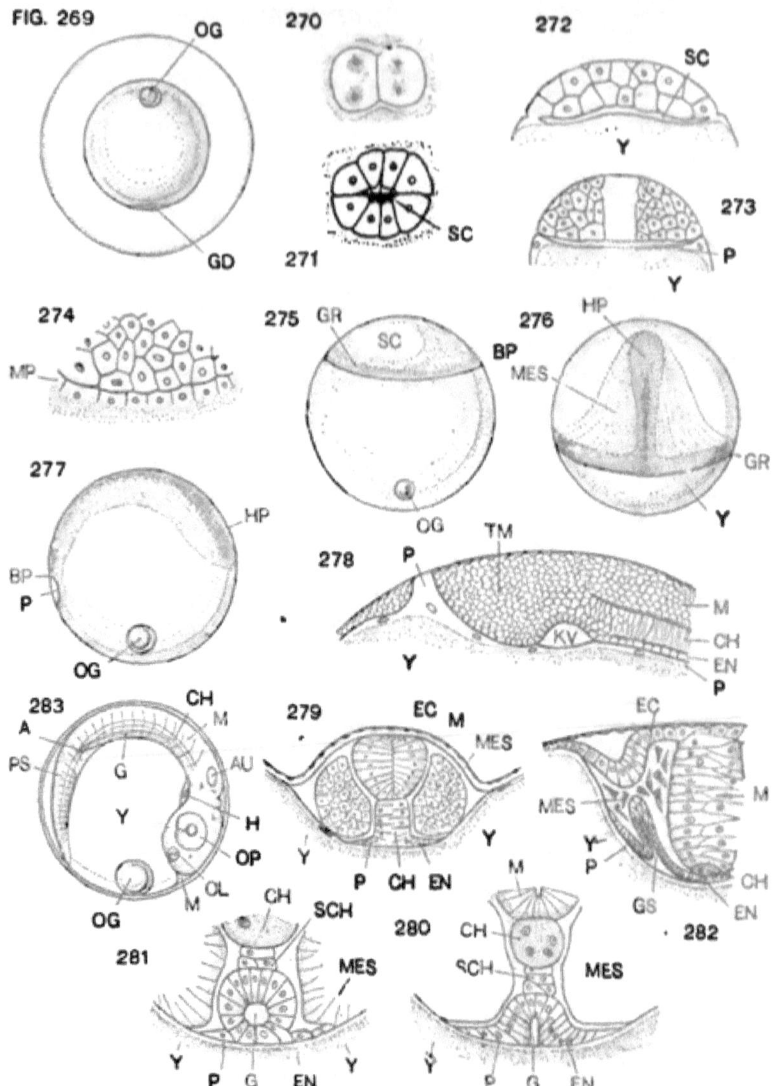

Figs. 269-283. — Development of Teleost, *Serranus atrarius.* (After H. V. Wilson.)
Fig. 276 × 25. 269. Egg immediately prior to segmentation, showing position of germ disc and of oil globule. 270. Germ disc after first cleavage. 271. Germ disc after third cleavage. 272. Vertical section of blastula. 273. Vertical section of blastula, showing origin of periblast. 274. View of marginal cells of blastula of similar stage. 275. Growth of blastoderm around yolk mass. 276. A slightly later stage, showing growth of embryo. 277. Continued growth of embryo and reduction in size of the blastopore. 278. Sagittal section of tail region of embryo of last figure. 279, 280, 281. Cross-sections of embryos, showing successive stages in the development of notochord, gut, neuron, mesoblast. 282. Cross-section of young embryo, showing the mode of formation of gill slit. 283. Embryo shortly before hatching.

A. Anus. *AU.* Auditory vesicle. *BP.* Dorsal lip of blastopore. *CH.* Notochord. *EC.* Ectoderm. *EN.* Entoderm. *G.* Gut. *GD.* Germ disc. *GR.* Germ ring. *GS.* Gill slit. *H.* Heart. *HP.* Head process. *KV.* Kupffer's vesicle. *M.* Spinal nervous system. *MES.* Mesoblast. *MP.* Marginal periblast cells. *OG.* Oil globule. *OL.* Olfactory pit. *OP.* Optic capsule. *P.* Periblast. *PS.* Primitive segments. *SC.* Segmentation cavity. *SCH.* Subnotochordal rod. *TM.* Tail mass. *Y.* Yolk.

V. *The Development of Teleost*

The mode of development of bony fishes differs in many and apparently important' regards from that of their nearest kindred, the **Ganoids.** In their **eggs a large** amount of yolk is present, and its relations to the embryo have become widely specialized.

As a rule, the egg of a Teleost is small, perfectly spherical, and enclosed in delicate but greatly distended **membranes** (Fig. 269). The germ disc, *GD*, **is especially** small, appearing on the surface **as an almost** transparent **fleck ;** it may occupy the same position **as in the other fishes,** or, as in **the figure, it may occur at the lowermost pole.** Among the fishes whose **eggs float at the surface** during development, **as of many** pelagic Teleosts, *e.g.* **the** Sea-bass, *Serranus* **atrarius,**—to **which all the** accompanying figures refer,—**the yolk is lighter** in specific gravity than the **germ; it is of fluid-like** consistency, almost transparent. **In the yolk at the upper pole of** the egg an oil globule, *OG*, **usually occurs ; this serves to** lighten the **gravity of** the entire egg, and from **its** position **must aid** materially in keeping **this pole of the** egg uppermost.

The early segmentation **of the** germ is seen **in Figs. 270, 271.** In the former, the first cleavage plane is **established,** and the nuclear divisions have taken place for the second; in the latter, the third cleavage **has been completed.** As in other fishes **these** cleavages are vertical, the third parallel to the first. **A** segmentation cavity, *SC*, occurs as a central **space between** the blastomeres, as it does in the sturgeon and gar-pike.

Stages **of** late segmentation are seen in section in Figs. **272, 273. In** both **the** segmentation cavity, *SC*, is greatly

P

flattened, but extends to the marginal cells of the germ disc; in Fig. 272 its roof consists of two tiers of blastomeres, its floor a thin film of the unsegmented substance of the germ; the marginal blastomeres are continuous with both roof and floor of the cavity, and are produced into a thin film which passes downward, around the sides of the yolk. In Fig. 273 the segmentation cavity is still further flattened; its roof is now a dome-shaped mass of blastomeres; the marginal cells have multiplied, and their nuclei are seen in the layer of the germ, *P*, below the plane of the segmentation cavity. These are seen at *MP* in the surface view of the marginal cells of this stage (Fig. 274); they are separated by cell walls only at the sides; below they are continuous in the superficial down-reaching layer of the germ. The marginal cells, *MP*, shortly lose all traces of having been separate; their nuclei, by continued division, spread into the layer of germ flooring the segmentation cavity, and into the delicate film of germ which now surrounds the entire yolk. Thus is formed the *periblast* of teleostean development, which from this point onward is to separate the embryo from the yolk; it is clearly the specialized inner part of the germ, which, becoming fluid-like, loses its cell walls, although retaining and multiplying its nuclei. It would accordingly correspond to that portion of the germ of the sturgeon in Fig. 253 which lies below the plane of the segmentation cavity, and which extends downward at the sides of the yolk; in this case, however, the surface outlines of the cells have not been lost. It will be seen from later figures (Figs. 278–282) that the periblast, *P*, comes into intimate relations with the growing embryo; it lies directly against it, and appears to receive cell increments from it at various regions; on the other hand, the nuclei of the periblast,

from their intimate relations with the yolk, are supposed
to subserve some function in its assimilation.

Aside from the question of periblast, the growth of
the blastoderm appears not unlike that of the sturgeon.
From the blastula stage of **Fig. 273 to** that of the early
gastrula (Fig. 275), the changes have **been** but slight; the
blastoderm has greatly flattened **out as its** margins grow
downward, leaving the **segmentation cavity** apparent at
SC. The rim of the blastoderm has become thickened,
as the 'germ ring;' and immediately in front of *BP*, **the**
dorsal lip of the blastopore, its thickening, **as in Fig. 255,**
marks the appearance of **the embryo. In Fig. 276** the
germ ring, *GR*, continues to **grow downward, and shows
more** prominently the **outline of the embryo; this now**
terminates at *HP*, **the head region; while on either ·side**
of this point spreads out tail-ward on **either side the indefi-**
nite layer of outgrowing **mesoderm, *MES*. In the stage**
of Fig. 277 the closure **of the blastopore, *BP*, is** rapidly
becoming completed; **in front of it stretches the widened**
and elongated form **of** the embryo. A sagittal **section**
through a late stage of the blastopore appears in Fig. 278;
with it **may** be compared the corresponding region **of the**
sturgeon **of** Fig. 256; the yolk plug, *YP*, of the **latter is**
now replaced by periblast, *P*, **the** dorsal lip at *BP*, **by**
TM, the tail mass, or more accurately the dorsal **section**
of· the germ rim; the cœlenteron under the **dorsal lip**
has here disappeared, on account of the close approximation of the embryo to the **periblast;** its **last** remnant,
the Kupffer's vesicle, ***KV*, is** shortly to disappear. **At**
TM, the germ layers become confluent as **at** *BP* in Fig.
256, but, unlike the sturgeon, the flattening **of the dorsal**
germ ring, *TM*, does not permit the formation of **a neu-**
renteric canal.

The process of the development of the germ layers in Teleosts appears an abbreviated one, although in many of its details it is but imperfectly known. In the development of the medullary groove, as an example, the following peculiarities exist : the medullary region at *HP* (Fig. 276) is but an insunken mass of cells without a trace of the groove-like surface indentation of Fig. 261 or 229. Its condition is figured at *M* in Fig. 282. It is only later, when becoming separate from the ectoderm, *EC*, that it acquires its rounded character (Fig. 279), *M*; its cellular elements then group themselves symmetrically with reference to a sagittal plane, where later by their disassociation (?) the canal of the spinal cord is formed (Fig. 280), *M*. The growth of the entoderm is another instance of specialized development. In the section of the embryo of Fig. 279, the entoderm exists in the axial region, its thickness tapering away abruptly on either side; its lower surface is closely apposed to the periblast; its dorsal thickening will shortly become separate as the notochord. In a following stage of development (Fig. 280), the entoderm is seen to arch upward in the median line as a preliminary stage in the formation of the cavity of the gut. Later, by the approximation of the entoderm cells in the median ventral line, the condition of Fig. 281 is reached, where the completed gut cavity exists at *G*.

The formation of the mesoderm in Teleosts is not definitely understood. It is usually said to arise as a process of 'delamination,' *i.e.* detaching itself in a mass from the entoderm. Its origin is, however, looked upon generally as of a specialized and secondary character.

The mode of formation of the gill slit of a Teleost does not differ from that in other groups; an evagination of the entoderm, *GS* (Fig. 282), coming in contact with an invaginated tract of ectoderm, *EC,* fuses, and at this point an opening is later established.

In Fig. 283 has been figured a late embryo. This may be compared with that of the sturgeon of Fig. 264. The Teleost, though of rounded form, is the more deeply implanted in the yolk sac; it is transparent, allowing notochord, primitive segments, heart, and sense organs to be readily distinguished; at about this stage both anus, *A,* and mouth, *M,* are making their appearance.

D. THE LARVAL DEVELOPMENT OF FISHES

When the young fish has freed itself from its egg membranes, it gives but little suggestion of its adult form. It enters upon a larval existence; which continues until maturity. The period of metamorphosis varies widely in the different groups of fishes—from a few weeks' to longer than a year's duration; and the extent of the changes that the larva undergoes are often surprisingly broad, investing every organ and tissue of the body,—the immature fish passing through a series of form stages which differ one from the other in a way strongly contrasting with the mode of growth of amniotes; since the chick, reptile, or mammal emerges from its embryonic membranes in nearly its adult form.

The fish may, in general, be said to begin its existence as a larva as soon as it emerges from its egg membranes. In some instances, however, it is difficult to decide at what point the larval stage is actually initiated: thus in sharks, the excessive amount of yolk material which has been provided for the growth of the larva renders unnecessary the emerging from the egg at an early stage; and the larval period is accordingly to be traced back to stages that are still enclosed in the egg membranes. In all cases the larval life may be said to begin when the following conditions have been fulfilled : the outward form of the larva must be well defined, separating it from the mass of yolk, its motions must be active, it must possess a continuous vertical fin fold passing dorsally from the head region to the body terminal, and thence ventrally as far as the yolk region ; and the following structures, characteristic in outward appearance, must also be established, the sense organs, — eye, ear and nose, — mouth and anus, and one or more gill clefts.

Among the different groups of fishes the larval changes are brought about in widely different ways. These larval peculiarities appear at first of far-reaching significance, but may ultimately be attributed, the writer believes, to changed environmental conditions, wherein one process may be lengthened, another shortened. So too the changes from one stage to another may occur with surprising abruptness. As a rule, it may be said the larval stage is of longest duration in (I) the Cyclostomes, and thence diminished in length in (II) Sharks, (III) Lung-fishes, (IV) Ganoids, and (V) Teleosts ; in the last-named group, a very much curtailed (*i.e.* precocious) larval life many often occur.

I. *Larval Cyclostomes*

The Cyclostome larva is represented in a stage as early as that of Fig. 212 : its form is here retort-shaped ; the yolk material is concentrated in the ventral region immediately in front of the blastopore (the anus ?), but is distributed in addition in the cells of other body regions. In the section of a slightly older larva (Fig. 215), in which the mouth is all but established, the form outline has become regular, the bulk of the yolk, *Y*, restricted to the cavity of the intestine, the only instance of this condition known among fishes (Cerátodus?), and, with but a single exception (Ichthyophis),* among all other vertebrates. The larval lamprey is by this time a quarter of an inch long, yellowish white in colour; its movements are sluggish, rarely more than to cause it to wriggle worm-like from the bottom. A few weeks later it has acquired its brownish grey colour, its fin fold is well marked, and its habit is active; it now feeds on muddy ooze rich in organic matter. It by this time possesses the essential characters of the well-grown larva, long looked upon as a distinct genus, *Ammocœtes*. In its larval stage the lamprey appears to live a number of years; in *Petromyzon planeri* the adult stage is said to be sometimes deferred until the autumn of the fourth or fifth year. The transformation is then a surprisingly sudden one; the head attains its enlarged size, the mouth its ring-like and suctorial character, losing its more anterior position, and its lip-like flaps (cf. Fig. 72, $C_1.D$); teeth are developed in place of the numerous mouth papillæ; gills, formerly simpler in character, opening directly from neck surface to gullet, now enter the branchial chamber, a ventral diverticulum of the gullet; eyes become prominent, complete their development, and attain the head surface; unpaired fin, formerly of great extent, is now reduced to its adult position and proportions.

II. *Larval Sharks*

The larval history of Sharks has been summarized in Figs. 284–289: the younger of these stages (Figs. 284, 285, 286) have not as yet escaped from their egg membranes. The hatching, in fact, of the young shark is

* The writer has not confirmed Salensky's observation upon the sturgeon.

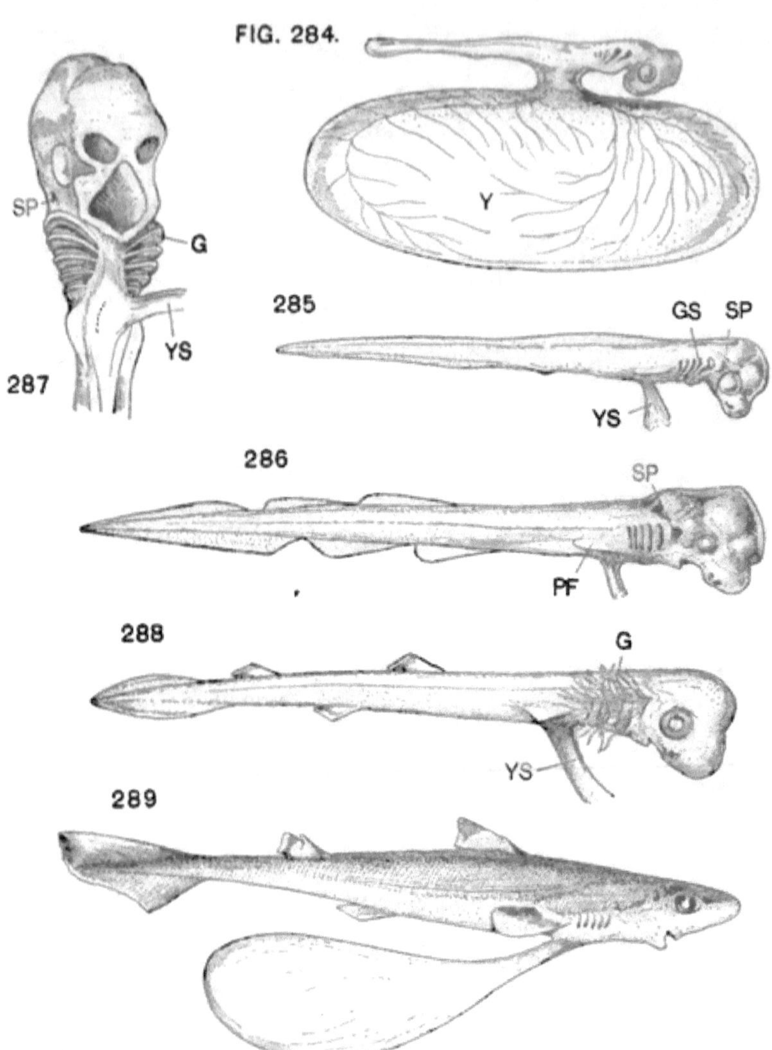

Figs. 284-289.—Larval sharks. (Figs. 284-287 after Balfour.) 284. *Pristiurus* (embryo, × 5) with yolk sac (× 2). 285, 286. Larvæ of *Scyllium*. × 4. 287. Ventral view of head of larval *Scyllium*, slightly younger than that of last figure. × 8. 288. Larva of *Acanthias*. × 4. 289. Late larva of *Acanthias*. × ¾.

G. Gills. *GS*. Gill slits. *PF*. Pectoral fin. *SP*. Spiracle. *Y*. Yolk sac. *YS*. Stalk of yolk sac.

an exceedingly slow one; Pristiurus emerges from the egg in about nine months, Scyllium in about seven. And in consequence of the large amount of yolk stored in the yolk sac, the young shark, as in Fig. 289, has fully acquired its adult outward characters by the time the yolk is exhausted and its sac absorbed.

In Fig. 284 is figured a stage in the development of Pristiurus which may be regarded as either embryonic or larval; the form of the larva is well established; gill clefts, muscle-plates, mouth, and sense organs are present; but, on the other hand, unpaired fin and anus are lacking. There is shown the abrupt constriction, characteristic of Elasmobranchs, which separates the animal from the yolk sac,—a construction which in later stages becomes narrow and tubular. The relatively larger size of the yolk sac in later stages is, of course, the result of the bulkier elaboration of the yolk material.

The youngest stage (Fig. 284) shows prominently the great enlargement of the anterior end of the embryo, a marked cephalic flexure, large optic capsule, and irregular gill slits of graded sizes; a tubular tail end, bulbous at the terminal, where the neurenteric canal occurs; as yet the nasal pits are in close proximity to the mouth. In the next stage (Fig. 285), the elongated trunk has its unpaired fin, the neurenteric canal disappearing; the beginnings of the pectoral fins are noticeable; gill clefts are of more uniform size; and the anal region is indicated. In the stage of Fig. 286, further advances are seen in the constricting off of the unpaired fins, the appearance of the ventral and the continued growth of the pectoral fins; in the reduced foremost gill slit (spiracle); in the jaw region, and, in fact, in the entire shaping of the head; in the appearance of the lateral line. In the ventral head region (Fig. 287), is to be noted the prominence of the mouth cavity, and the enlarged gill arches, showing by this time the outbudding branchial filaments. In the stage of Fig. 288, the larva begins to appear shark-like; the fins are longer and more noticeable, the anus has appeared, and the branchial filaments by continued growth protrude at all gill openings. The external gills thus acquired are seen in a later stage (Fig. 289) to have disappeared; they have aided, however, as Beard, Turner, and others have shown, in absorbing nutriment, and must be looked upon as an especial organ of the larval life of the animal. Fig. 289 illustrates a final larval stage: in it there appear all of the structures of the adult outward form, *e.g.* shagreen, fin spines, nictitating membrane, anterior and posterior nasal openings. This larva has been estimated to be about a year older than that of Fig. 284.

III. *Larval Lung-fish*

The larval history of the lung-fish, Ceratodus, as recently described by Semon, seems to offer characters of exceptional interest, uniting features of Ganoids with those of Cyclostomes and Amphibians.

The newly hatched Ceratodus (Fig. 290) does not strikingly resemble the early larva of shark (Fig. 284). No yolk sac occurs, and the distribution of the yolk material in the ventral and especially the hinder ventral region is suggestive rather of lamprey or amphibian; it is, in fact, as though the quantum of yolk material had been so reduced that the body form had not been constricted off from it. The caudal tip in this stage appears, however, to resemble that of the shark, and as far as can be inferred from surface views a neurenteric canal persists. Like the shark there then exists no unpaired fin; the

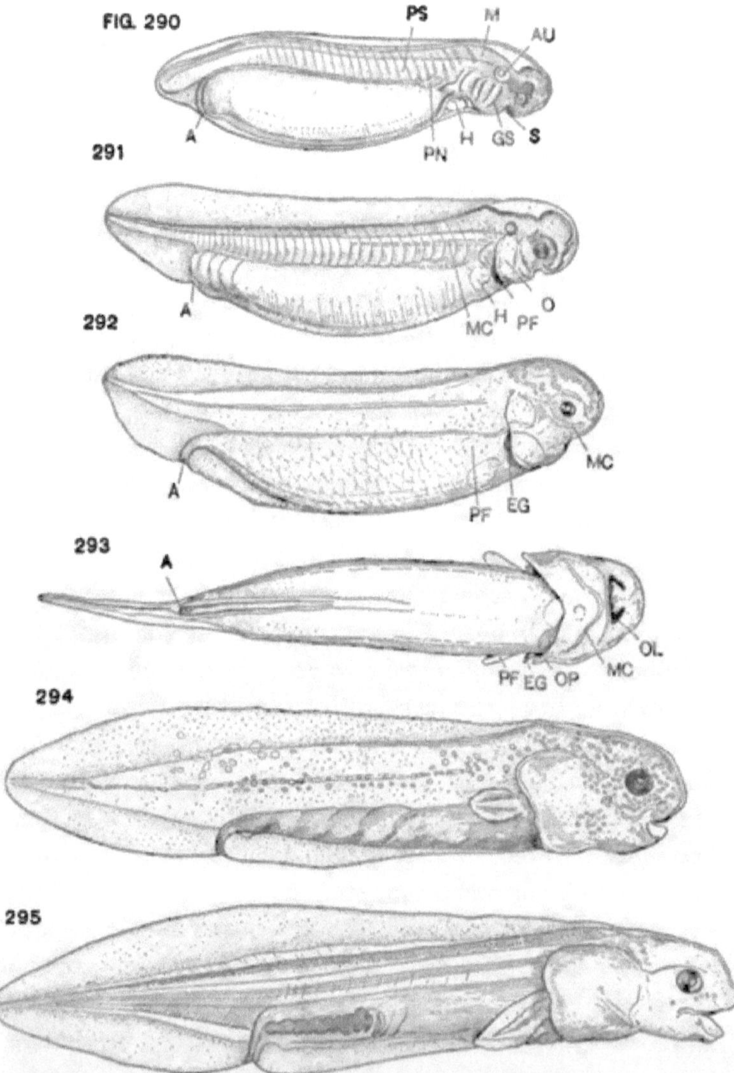

Figs. 290-295. — Larval lung-fishes, *Ceratodus*. (After SEMON.) × 6. 290. Embryo at about the time of hatching. 291. Young larva. 292. Larva of two weeks. 293. Larva of four weeks, ventral side. 294. Larva of six weeks. 295. Larva of ten weeks.

A. Anus. *AU.* Auditory vesicle. *EG.* External gills. *GS.* Gill slits. *H.* Heart. *M.* Central nervous system. *MC.* Mucous canals. *O.* Opercular flap. *OL.* Olfactory organ. *PF.* Pectoral fin. *PN.* Pronephros. *PS.* Primitive segments. *S.* Mouth pit, stomodæum.

gill slits, five (?), *GS*, are well separated, and there is an abrupt cephalic flexure. In this stage pronephros and primitive segments, *PS*, are well marked, and are outwardly similar to those structures in Ganoid ; the mouth, *S*, is on the point of forming its connection with the digestive cavity ; the anus is the persistent blastopore ; the heart, well established, takes a position, as in Cyclostomes, immediately in front of the yolk material.

In a later stage the unpaired fin has become perfectly established, the tail increasing in length ; the gill slits have now been almost entirely concealed by a surrounding dermal outgrowth, the embryonic operculum ; a trace of the pectoral fin, *PF*, appears ; the lateral line is seen proceeding down the side of the body ; near the anal region the intestine * becomes narrower and the beginnings of the spiral valve appear. In a larva of two weeks (Fig. 292), a number of developmental advances are noticed : the fish has become opaque, the primitive segments are no longer seen ; the size of the yolk mass is reduced ; the anal fin fold appears ; sensory canals are prominent in the head region ; lateral line is completely established ; the rectum becomes narrowed ; and the cycloidal body scales are already outlined. Gill filaments may still be seen beyond the rim of the outgrowing operculum. In the ventral view of a somewhat later larva (Fig. 293), the following structures are to be noted : the pectoral fins which have now suddenly budded out,† reminding one in their late appearance of the mode of

---

* The yolk appears to be contained in the digestive cavity as in Ichthyophis and lamprey.

† The abbreviated mode of development of the fins is most interesting ; from the earliest stage they assume outwardly the archipterygial form ; the retarded development of the limbs seems curiously amphibian-like ; the pectorals do not properly appear until about the third week, the ventrals not until after the tenth.

origin of the anterior extremity of urodele ; **the greatly en-**
larged size **of** the opercular **flap ;** external gills, **still** prominent ; the internal nares, *OL,* becoming constricted off **into**
the mouth cavity by the dermal fold **of the anterior lip (as**
in some sharks), and finally **(as** in Protopterus and **some**
batrachian larvæ) **the one-sided** position of the anus.

The larva of six **weeks** (Fig. 294) suggests the **outline**
of the mature fish ; head and sides show the various openings of the tubules **of** the insunken sensory canals ; and
the 'archipterygium' of the pectoral fin is **well defined.**
The oldest larva figured (Fig. **295) is ten weeks old ; its**
operculum and pectoral **fin show an increased size ; the**
tubular mucous **openings, becoming finely subdivided, are**
no longer noticeable ; and although **the basal supports of**
the remaining fins **are coming to be established, there is**
as yet little more than **a trace of the ventrals.**

IV. *Larval Ganoids*

The larval forms **of a Ganoid,** Acipenser (Figs. **296–**
302), resemble far **more closely those of** the shark **than of**
the lung-fish. When newly hatched, the young sturgeon
(Figs. 296, 297) **is** attached **to** the well-rounded **yolk sac**
situated in the throat region, in exactly the **position one**
would expect **the yolk** stalk to be situated if the yolk mass
were larger ; it resembles **the** shark **larva** of **Fig.** 295 in
its unpaired fin, in gill slits, in olfactory, *OL,* optic, *OP,*
and auditory, *AU,* organs, and in the fact that it possesses
even at this stage a trace of **the neurenteric canal ;** on the
other hand, it suggests the **Ceratodus larva of Fig. 291 in**
its stout **trunk region, prominent muscle segments,** pronephros, *PN,* and **anus,** *A* ; **at** the **foremost corner of** the
yolk sac are mouth pit (stomodæum, *S*) **and** heart. A
larva **of** the second day resembles **in** many features the

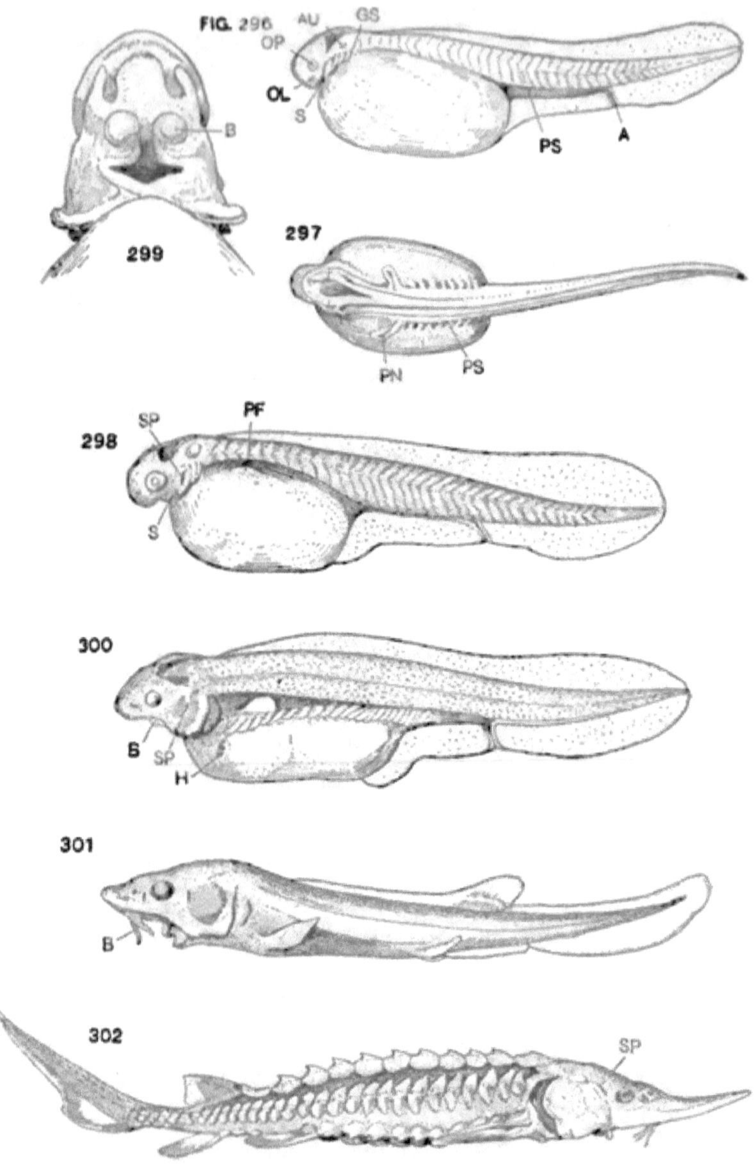

Figs. 296–302. — Larval sturgeons. (All but Fig. 302 after KUPFFER.) Fig. 299, × 18; 296–300, × 10; 301, × 8; 302, × ⅓. (Enlargement approximate.) 296, 297. Larvæ shortly after hatching. 298. Larva two days old. 299. Mouth region of larva of third day. 300. Larva of fourth day. 301. Larva of twenty-eight days. 302. Sturgeon of twelve months.

A. Anus. AU. Auditory vesicle. B. Barbel. GS. Gill slit. H. Heart. OL. Olfactory pit. OP. Optic vesicle. PF. Pectoral fin. PN. Pronephros. S. Mouth pit. SP. Spiracle.

shark larva of Fig. 286: **dorsal,** caudal, and anal regions are outlined in the unpaired fin; **a pectoral fin of a fin-fold** character, *PF,* has appeared; **the spiracle, *SP*, is becom-**ing established. The mouth **region is more clearly indi-**cated in this stage, *S,* but **may better be seen in ventral** view in a slightly later larva; here (Fig. 299) **the posterior** lip is constricted **off from** the yolk region, and the **anterior** lip is budding off near the median **line** a pair of **the** tactile barbels; the dermal fold (operculum) enclosing **the** gills **is** in a condition very similar to that **of** Ceratodus **in** Fig. 293. **A** larva of **the fourth day (Fig. 300)** shows well-marked advances : **the snout is elongated ; the** opercle is enclosing the gills, which are **now seen to protrude as** external branchial ; **the pectoral fin elongates and is tend-**ing to protrude **its fin axis; body segments and heart are** encroaching into the region **of the now elongate** yolk sac; the lateral line has been formed. In a larva of four **weeks** (Fig. 301), the essential outlines of the sturgeon **may be** recognized, although the head appears of strikingly larger proportions: barbels, **nares, mouth,** operculum, and spiracle **are as** in the adult ; **fins,** of the mature outlines, are want-ing in all save basal supports ; yolk material has **long** since been exhausted. **A** very late larva (Fig. 302), supposed **to** be twelve months old, differs outwardly from the sexually mature **form** in **but its** colouring and dermal **plates** : those of the regular rows are of great size, conspicuous in their abrupt spines and well-roughened borders ; and those of the remaining trunk integument are remarkably prominent ; the tail of the larva shows clearly its palæoniscoid character.

V. *Larval Teleosts*

The metamorphoses of the **newly** hatched Teleost must finally be reviewed ; they **are certainly** the **most**

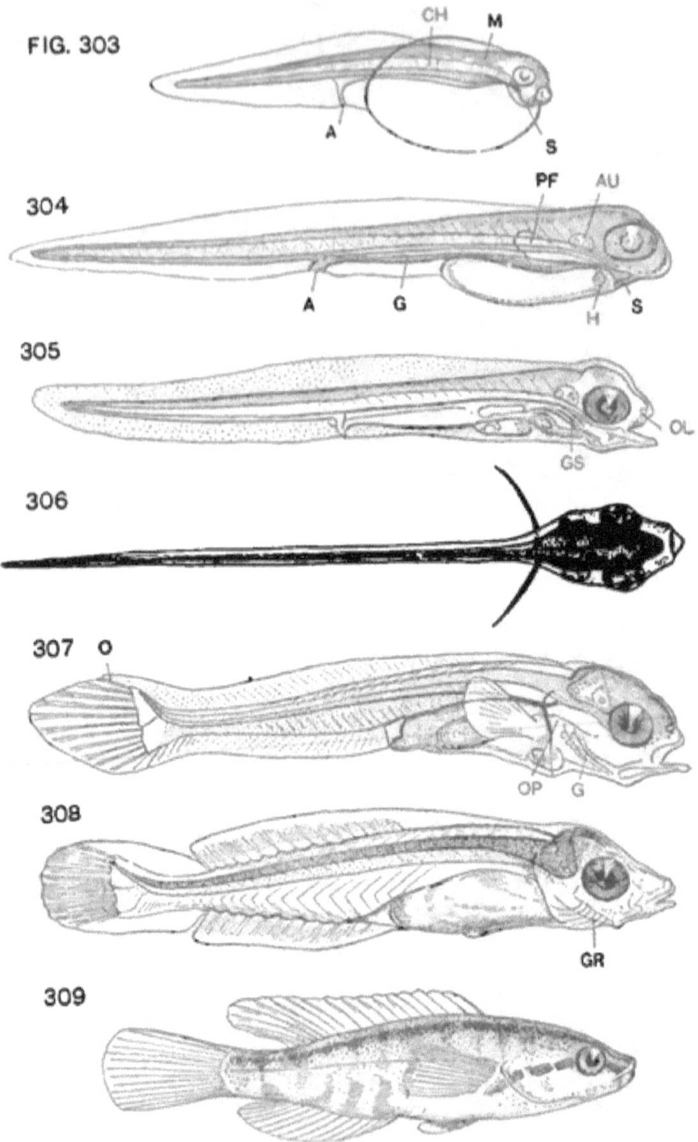

Figs. 303-309.—Larvæ of Teleost, *Ctenolabrus*. (After A. AGASSIZ.) Fig. 309 × about 7, other figures × about 14. 303. Larva shortly after hatching. 304, 305. Larvæ of first few days. 306, 307. Larva of one week. 308. Larva of two weeks (?). 309. Final larval stage, four (?) weeks.

A. Anus. *A U.* Auditory vesicle. *CH.* Notochord. *GR.* Gill protecting dermal rays. *H.* Heart. *M.* Central nervous system. *OL.* Olfactory capsule. *OP.* Optic vesicle. *PF.* Pectoral fin. *S.* Stomodæum.

varied and striking of all larval fishes, and, singularly
enough, appear to be crowded into the briefest space of
time; the young fish, hatched often as early as on the
fourth day, is then of the most immature character; it
is transparent, delicate, inactive, easily injured; within
a month, however, it may have assumed almost every
detail of its mature form. A form hatching three mille-
metres in length may acquire the adult form before it
becomes much longer than a centimetre.

The larval life of the common Sea-bream, or Cunner,
Ctenolabrus cœruleus, has been admirably figured by A.
Agassiz. The newly hatched fish (Fig. 303) has the yolk
sac appended at the throat, as a large, transparent, if
slightly tinted, globule; save for its great delicacy and
transparency, it may generally be compared to the corresponding larva of Acipenser (Fig. 296). By the third day
(Fig. 304), the yolk sac has become greatly reduced, the
trunk elongated, the fin fold less conspicuous; primitive
segments have appeared; the pectoral fin has arisen, but
is not of the elasmobranch form of the similar stage (Fig.
298) of sturgeon; it is long, thin, transparent, and its
rapid growth indicates its metamorphosed character. The
mouth, S, is in this stage on the point of formation. In
a slightly older larva (Fig. 305), the yolk has almost disappeared, its gill slits, GS, and mouth have now been
formed, and with the latter the nasal apertures. In a following stage (Figs. 306, 307), a well-marked opercular fold
makes its appearance; pectoral fins acquire their completed outline and the fin fold undergoes changes: anteriorly it acquires supporting actinotrichia, posteriorly the
dermal supports of the caudal fin appear and at their bases
the coalesced radio-basals; a ganoidean heterocercy is
here apparent, its distal tip the membranous opisthure, O.

The later larva (Fig. 308) is characterized by the appearance of abundant pigment masses (not shown in the figure) in all regions of the trunk; branchiostegal rays, *GR*, and traces of pelvic fins are noted; the caudal fin has become separated from the dorsal and anal elements. And finally, in the stage of Fig. 309, the fish, although still of very small size, has acquired almost perfectly its mature features; the outward differences are only those of pigmentation and fin proportions.

Acanthodes, ἀκανθώδης, provided with **spines.**
Acanthopterygii, ἄκανθα, spine, πτέρυξ, fin(ned).
Acipenser, ἀκιπήσιος, classic name of sturgeon.
Actinopterygii, ἀκτίς, stout ray, πτέρυξ, **fin(ned).**
Alopias, ἀλωπεκίας, classic name of the **fox shark.**
Amia, ἀμία, classic name of **tunny(?).**
Amiurus, ἀμία, Amia, οὐρά, **tail(ed).**
Ammocœtes, ἄμμος, sand, κοίτη, (a bed) abider.
Anacanthini, ἀνά, without, ἄκανθα, spine.
Anguilla, classic name of eel.
Arthrodira, ἄρθρον, joint, (?)δίς, double.
Aspidorhynchus, ἀσπίς, shield, ῥύγχος, snout.

Bdellostoma, βδέλλα, leech, στόμα, **mouth.**
Belonorhynchus, βελόνη, **classic name** of gar-fish, ῥύγχος, snout.

Calamoichthys, *calamus*, **a reed,** ἰχθύς, **fish.**
Callichthys, κάλλος, beautiful, ἰχθύς, fish.
Callorhynchus, κάλλος, beautiful, ῥύγχος, **snout.**
Carassius, χάραξ, classic **name** of (sea)fish.
Caturus, κατά, on the under **side,** οὐρά, **tail.**
Cephalaspis, κεφαλή, **head,** ἀσπίς, shield.
Ceratodus, κέρας, horn, ὀδούς, tooth(ed).
Cestracion, κέστρα, classic name of (pavement-toothed) sea-fish.
Cheirodus, χείρ, hand, ὀδούς, tooth(ed).
Chimæra, χίμαιρα, fabulous **monster,**—lion's head, goat's body, dragon's
 tail.
Chlamydoselache, χλαμυδός, frilled, σελάχη, shark.
Chondrostei, χόνδρος, cartilage, ὀστέον, bone(d).
Cladoselache, for Cladodonto-selache, κλάδος, branch, ὀδούς, tooth(ed),
 σελάχη, shark.
Climatius, κλίμα, **a gradation** (in allusion, perhaps, to the graded row
 of fin spines).

Coccosteus, κόκκος, rough like a berry, ὀστέον, **bone.**
Cœlacanthus, κοῖλος, hollow, ἄκανθα, spine(d).
Crossopterygii, κροσσός, fringe or tassel, πτέρυξ, **fin.**
Ctenodus, κτείς (κτενός), comb, ὀδούς, tooth(ed).
Cyclostomata, κύκλος, circular, στόμα, **mouth.**

Dinichthys, δεινός, terrible, ἰχθύς, **fish.**
Diplognathus, διπλός, double (pointed), **γνάθος, jaw.**
Diplurus, διπλός, double, οὐρά, tail(ed).
Dipnoi, δίπνοος, double breathing.
Dipterus, δίς, two, πτέρον, fin(ned).

Edestus, **ἐδεστής, a devourer.**
Elasmobranchii, **ἐλασμός,** strap-like, **βράγχια,** gill(ed).
Elonichthys, (?)ἐλύω, to twist, ἰχθύς, fish.
Erythrinus, **ἐριθρός, red-coloured.**
Eurynotus, **εὐρύς,** wide, νῶτος, **back(ed).**
Eusthenopteron, **εὐσθενής, strong, πτερόν, fin.**

Fierasfer, derivation of Cuvier uncertain, perhaps from proper name.

Gadus, **classic name of cod.**
Ganoid, **γάνος,** enamelled.
Gnathostome, γνάθος, jaw, στόμα, mouth.
Gyroptychius, γῦρος, a circle, πτύχιος, folded (referring **to** the tooth
    enamel).

Harriotta, from **the** proper name Harriott.
Hemitripterus, *hemi,* half, τρεῖς, three, πτερόν, fin(ned).
Heptanchus, ἑπτά, **seven, ἄγχω** (referring **to** the compressed gill
    openings).
Hippocampus, classic name, "sea-horse."
Holocephali, ὅλος, whole or complete, κεφαλή, head.
Holoptychius, ὅλος, entire(ly), πτύχιος, folded (referring to the tooth
    enamel).
Hybodus, ὗβος, **hump,** ὀδούς. **tooth.**
Hyperoartia, ὑπερῴα, palate, ἄρτιος, **entire.**
Hyperotretia, ὑπερῴα, palate, τρετός, pierced.

Ichthyotomi, ἰχθύς, **fish, τέμνω,** separate (referring **perhaps** to the
    distinctness of this group).
Ischyodus, ἰσχύς, power(ful), ὀδούς, tooth(ed).

Læmargus, classic name of a shark.
Lagocephalus, λαγώς, rabbit, κεφαλή, head.
Lamna, λάμνα, classic name for a shark.
Lepidosiren, λεπίς, scale(d), *siren*, **salamander.**
Lepidosteus, λεπίς, scale, ὀστέον, bone.
Leptolepis, λεπτός, smooth or delicate, **λεπίς, scale(d).**
Lophobranchii, λόφος, tuft, βράγχιον, gill(ed).

Marsipobranchii, μαρσίπιον, pouch, βράγχια, gills.
Megalurus, μέγας, large, οὐρά, tail(ed).
Microdon, μίκρος, small, ὀδούς, tooth(ed).
Mormyrus, classic name of a (sea) fish (— from μορμύρω, I murmur).
Myliobatis, μυλίας, pavement (toothed), **βατίς, skate.**
Mylostoma, μύλος, mill(like), στόμα, mouth.
Myriacanthus, μυριάς, ten thousand, ἄκανθα, spine.
Myxine, μυξῖνος, slimy-fish.

Onychodus, ὄνυξ, claw ; **ὀδούς, tooth(ed).**
Ophidium, ὀφίδιον, a snake.
Osteolepis, ὀστέον, bone, λεπίς, scale(d).
Ostracoderm, ὀστράκιον, **shell, δέρμα, skin.**

Palæaspis, παλαιός, ancient, **ἄσπις, shield.**
Palæoniscus, παλαιός, ancient, ὀνίσκος, a **sea-fish.**
Palæospondylus, παλαιός, ancient, σπόνδυλος, vertebræ.
Parexus,? παρέχω, have as one's own (referring to the peculiar **nature**
    of the fish?)
Perca, classic name of fish.
Petromyzon, πέτρος, stone, μυζάω, to suck.
Phaneropleuron, φανερός, well marked, πλευρά, side (fins) or ribs(?).
Pisces, fishes.
Plagiostomi, πλάγιος, transverse, στόμα, mouth.
Plectognathi, πλεκτός, twisted, γνάθος, jaw.
Pleuracanthus, πλευρά, side, ἄκανθα, spine.
Pleuropterygii, πλευρά, side, πτέρυξ, fin(ned).
Pogonias, πωγωνίας, bearded.
Polyodon, πολύς, **many, ὀδών, tooth(ed).**
Polypterus, πολύς, **many, πτερόν, fin(ned).**
Prionotus, πρίων, saw, **νῶτος, back.**
Pristiophorus, πρίστις, a saw, φορέω, **to carry.**
Pristis, **πρίστις, a** saw-fish.
Protopterus, πρῶτος, ancient, πτερόν, fin(ned).

Psammodus, ψάμμος, sand, ὀδούς, tooth(ed).
Psephurus, ψῆφος, a little stone, οὐρά, tail.
Pseudopleuronectes, ψεῦδος, false, πλευρόν, side, νήκτης, swimmer.
Pterichthys, πτέρυξ, fin or wing, ἰχθύς, fish.

Raja, classic name of skate.
Rhabdolepis, ῥάβδος, nail, λέπις, scale(d).
Rhina, ῥίνη, a rasp.
Rhinobatus, ῥίνα, Rhina, βατίς, skate.
Rhynchodus, ῥύγχος, snout, ὀδούς, tooth(ed).

Scaphirhynchus, σκαφίον, shovel, ῥύγχος, snout.
Scomberomorus, σκόμβρος, mackerel, μόριον, part.
Scyllium, σκύλιον, classic name of this shark.
Selachii, σελάχη, shark.
Semionotus, σημεῖον, a standard, νῶτος, back.
Silurus, classic name of fish.
Siphostoma, σίφων, tube, στόμα, mouth.
Sirenoidei, *siren*, salamander, οἶδος, like.
Squaloraja, *squalus*, shark, *raja*, skate.
Squalus, classic name of a shark.
Squatina, a classic name of a sea-fish.

Teleocephali, τέλεος, entirely, ὀστέον, bone, κεφαλή, head.
Teleost, τέλεος, entirely, ὀστέον, bone.
Teleostomi, τέλεος, entirely, ὀστέον, bone, στόμα, mouth.
Titanichthys, *titan*, giant, ἰχθύς, fish.
Torpedo, classic name (from the root of Torpor, stupefy).
Trachosteus, τραχύς, rough, ὀστέον, bone.
Trygon, τρύγων, the thorny ray.

Urogymnus, οὐρά, tail, γύμνος, naked.

Xenacanthus, ξένος, strange, ἄκανθα, spine.

BIBLIOGRAPHY

In the following list the **writer** aims to present **the** more recent and more important works relating to the general subject **of fishes.** Titles have been classified, and most of the references give more or less complete bibliographies of their special subjects. Of the journals in which papers occur the principal abbreviations **are** as follows : —

A . . . Archiv (or Archives).	Q.J.M.S	Quarterly Journal **of** Microscopical Science.
AH . . Abhandlungen.		
Ann. N.H Annals and Magazine of Natural History.	P . . . Proceedings.	
	R . . . Report.	
B . . . Bulletin.	S . . . Society.	
C.R . . Contes rendus.	SB . . Sitzungsberichte.	
DS . . . Denkschriften.	Sci . . Science, **or Scientific.**	
J . . . Journal	Tr. . . Transactions.	
JB . . . Jahrbuch	U.S.F.C United States Fishery Commission.	
JH . . . Jahreshefte		
J.R.M.S . Journal of the Royal Microscopical Society.	VH . . Verhandlungen.	
	Z . . . Zeitschrift.	
MT . . Mittheilungen		

The Roman numerals denote the number of the volume, the Arabic numerals the pages.

WORKS ON **THE** GENERAL SUBJECT, FISHES

WOODWARD, A. SMITH Catalogue of Fossil Fishes in the British Museum. Vols. I, II (and III).
London, 1889–(95).

GÜNTHER, A. . . . Catalogue of the Fishes in the British Museum. Vols. I–VIII. London, 1859–70.

GÜNTHER, A. . . . An Introduction to the Study of Fishes. 8vo. pp. 720. Illustrated. Edinburgh, 1880.

GÜNTHER, A. . . . Fishes: Challenger Reports. Vol. I, pt. VI, Vol. XXXI, pt. LXXVIII.
London, 1880–89.

GILL, T. . . . . . Fishes: Standard Natural History.
Boston, 1885.

<table>
<tr><td>GOODE, G. BROWN . .</td><td>Fishery Industries of U. S. U. S. F. C.
Washington, 1884.</td></tr>
<tr><td>DUMERIL, A. . . . .</td><td>Histoire naturelle des Poissons. Vols.
I–II (Sharks, Chimæroids, Lung-fishes,
Ganoids, Lophobranchs).   Paris, 1890.</td></tr>
<tr><td>AGASSIZ, L. . . . .</td><td>Recherches sur les Poissons Fossiles. Vols.
I–V, with Atlas volumes.
Neuchâtel, 1833–43.</td></tr>
<tr><td>ZITTEL, K. v. . . .</td><td>Handbuch der Palaeontologie. Fische.
Munich, 1887.</td></tr>
<tr><td>ROLLESTON, G. . . .</td><td>Forms of Animal Life. Second edition.
Oxford, 1888.</td></tr>
<tr><td>HUXLEY, T. . . . .</td><td>Manual of the Comparative Anatomy of
Vertebrated Animals.   New York, 1872.</td></tr>
<tr><td>JORDAN and GUILBERT</td><td>Manual of the Vertebrates of Eastern N. A.
McClurg. Last edition.</td></tr>
</table>

SKELETON. — '86 BAUR, G., Squamosum, Anat. Anz.   '87 Ribs,
Am. Nat. xxi, 942–945. '86 COPE, E. D., Caudal vertebræ, Am.
Phil. Soc. 243.   '93 BOULENGER, G. A., Hæmapophyses, Ann.
N.H. xii, 60–61. '92 DOLLO, L., Ribs, vertebræ, B. Sci. Fr. Belg.
xxiv. '87 GEGENBAUR, Occipital region, Kölliker Festschr. 1–33.
'79 GOETTE, A., Wirbelsäule, A. mikr. Anat. xvi, 428. '89 HAT-
SCHEK, Rippen, VH. Anat. Gesell. Berl. (Jena). '78 IHERING,
H., Wirbelverdoppelung, Zool. Anz. I, 72–74. '93 JORDAN, D. S.,
Temperature and vertebræ, Wilder Quarter Century Book, Ithaca,
13–37.   '93 KLAATSCH, H. (Vertebræ), Morph. JB. xix, 649–
680, and xx, 143–186. '68 KLEIN, Schädel, Würt. Nat. JH. 71–
171, and ('81) xxxvii, 326–360.   '87 LUOFF, B. (Chorda and
Sheath), B. S. Mosc. 227–342 (442–482, German).   '77 PARKER
and BETTANY, Morph. of the Skull, London, pp. 14–90.   '89
POUCHET and BEAUREGARD, Traité de Ostéol. Comp. Paris,
398–451. '87 STRECKER, C. (Condyles), A. Anat. Phys. Anat.
Abth. 301–338.

INTEGUMENT, TEETH. — '92 AGASSIZ, A., Chromatophores, B.
Mus. Comp. Zool. xxiii, 189–193.   '82 BAUME, A., Odont. Forsch.
Leip. 41–52.   '77 HERTWIG, O., Hautskelet, Morph. JB. II,
328–395, and v ('79), 1–21. '90 KLAATSCH, H., Schuppen, op.
cit., 97–202 and 209–258. '45 OWEN, Odontography, London.
'93 RYDER, J. A., Mechanical genesis of Scales, Ann. N. H., xi,
243–248. '82 TOMES, C., Dental Anat. Ed. 2.

FINS. — '90 COPE, Homologies, **Am. Nat.** 401–423. **'79** DAVIDOFF, M., Pelvics, Morph. JB. v, 450–520, vi ('80), 125–128, **433–468.** '87 EMERY, C., Homologies, Zool. **Anz. x, 185–189.** **'65** GEGENBAUR, C., Brust Flosse, Leip. 4to, pp. 176. **'70** Jen. Z., v, and ('73) Archipterygium, **vii.** '79 Morph. JB., v, **521–525.** '94 Op. cit. xxi, 119–160. **'89** HATSCHEK (Paired), VH. Anat. Gesell. Berl. 82–90. **'68** PARKER, W. K., Shoulder girdle, Ray Society, Lond. **pp. 237.** **'83** RAUTENFELD, E. V., Ventrals, Dorpat ('82), 48 pp. **'79** RYDER, J. A., Bilateral symmetry, **Am. Nat.** xiii, 41–43. **'85** Unpaired fins, op. **cit.** xix, 90–97. **'86** Embryol. of fins, R. **U. S. F. C., 981–1086.** '86 Fin rays and degeneration, P. **U. S. Nat. Mus. 71–82.** '87 Homologies, P. Acad. Philadel. 344–368. **'77** THACHER, J., Homologies, Tr. Conn. Acad. III. '92 WIEDERSHEIM, R., Gliedmassenskelet, Jena, **266 pp.** **'92** WOODWARD, A. S., Evolution, **Nat. Sci. 28–35.**

VISCERA, GLANDS, CIRCULATORY. — '84 AYERS, H., Pori abdominales, **Morph. JB. x, 344–349.** '89 Carotids, B. Mus. **Comp.** Zool. xvii. **'82** BALFOUR, F. M., Head kidney, **Q. J. M. S. xxx, 12–16.** '87 BOAS, J. E. V., Arterienbogen, **Morph. JB. xiii,** 115–118. '79 BRIDGE, **T., Pori abdominales, J. Anat.** Phys. **xiv,** 81–102. '85 CLELAND, **J., Spiracle, R. Br. Ass.** 1069. **'87** EBERTH, C. J., Blutplättchen, Kölliker **Festschrift, 37–48.** **'66** GEGENBAUR, Bulbus, **Jen. Z. ii, 365–375.** '84 Abdominal poren, Morph. JB. x, 462–464. **'91** Conus, op. cit. xvii, 596, 610. '85 GROSGLIK, S., Kopfniere, **Zool. Anz. viii, 605–611.** '90 HOWES, G. B., Intestinal **canal and blood supply, J. Linn.** S. xxiii, 381–410. '64 HYRTL, **J. (Hepatic and portal), SB. Acad. Wiss. Wien,** 167–175. '85 PHISALIX, C., Rate, **Paris, 8vo.** '90 RÖSE, C., Herz, Morph. JB. **xvi, 27–96.** '82 SOLGER, B., Niere, A. H. Gesell. Halle, **xv, 405–444.** '84 WELDON, W. F. R., Suprarenals, P. Roy. S. xxxvii, 422–425.

SWIM-BLADDER. — '86 ALBRECHT, P., Non-homologie des poumons, Paris and Brux. 44 pp. **'80** DAY, F., Zool. 97–104. '66 GOURIET, E., Ann. Sci. Nat. vi, 369–382. **'73** HASSE, C., Anat. Studien, I, Heft **4.** '90 LIEBREICH, O., A. Anat. Phys. Phys. Suppt. 142–161, 360–363. '85 MORRIS, C., P. Acad. Nat. Sci. Philadel. 124–135, Anat. Anz. ('85) xxvi, 975–986.

NERVOUS SYSTEM AND END ORGANS. — '83 BAUDELOT, E., fol. Paris, 178 pp. '88 BATESON, Sense organs, J. Mar. Biol. Ass. I, No. 2. '85 BEARD, J., Branchial sense organs, Q. J. M. S. xxvi. '82 BERGER, E. (Eye), Morph. JB. viii, 97–168. '84 BLAUE, J. (Nasal membrane), A. Anat. Phys. 331–362. '83

CANESTRINI, Otoliths, Atti. Soc. Pad. viii, 280–339. '86 Hearing organ, op. cit. ix, 256–282. '91 CHEVREL, R., Sympathetic, Thèse faculté des sciences, Paris. '79 DERCUM, F., Lateral line, P. Acad. Phil. 152–154. '70 FÉE, F., Système lateral, Mem. S. Sci. Nat. Strasb. vi, 129–201. '73 HASSE, C., Gehörorgan, Anat. Stud. I, Heft 3. '88 JULIN, C., Epiphysis, B. Sci. Nord. x, 55–65. '90 ? KOKEN, E., Otoliths, Z. geol. Gesell. xliii, 154. '91 OWSJANNIKOW, P. (Pineal eye), Rev. S. Nat. St. Petersb. 100–111. '81 RETZIUS, G., Gehörorgan, Stockholm, fol. 222 pp. '71 SCHULTZE, F. E., Seitenlinie, A. mikr. Anat. vi, 62. '70 STIEDA, L., Centralnervensystem, Z. wiss. Zool. xxi, 273–456.

EMBRYOLOGY.—'85 HAACKE, W., Uterinaler Brutpflege, Zool. Anz. viii, 488–490. HALBERTSMA, H. J., Normal en abnormal Hermaphroditismus, Tijd. Nied. Dier. Ver. Amst. '87 HOCH-STETTER, F., Venensystem, Morph. JB. xiii, 119–172. '86 HOFF-MAN, C. K., Urogenital, Z. wiss. Zool. xliv, 570–643. '91 KUPFFER, C. v., Kopfniere, VH. Anat. Gesell. 22–55. '90 LAGUESSE, E., Rate, J. de l'Anat. Phys. xxvi, 345–406 and 425–495. '77 LANKESTER, E. RAY. Germ layers, Q. J. M. S. xvii. '93 LWOFF, B., Keimblätterbildung, Biol. Centralb. xiii, 40–50, 76–81. '79 MAR-TENS, E. V., Hermaphroditische Fische, Naturf. 116. '80 NUSS-BAUM, M., Differenzirung d. Geschlechts, A. mikr. Anat. xviii, 1–121. '89 SCHWARZ, D., Schwanzende, Z. wiss. Zool. xlix, 191–223. '92 VIRCHOW, H., Dotterorgan, Z. wiss. Zool. liii, Suppl. 161–206.

THE CYCLOSTOMES

GENERAL.—'93 AYERS, H., Bdellostoma, Woods Holl Lectures, 125–161. '92 BEARD, J., Lampreys and Hags, Anat. Anz. viii, 59–60. '91 BUJOR, P., La metamorphose de l'Ammocœtes, Rev. Biol. du Nord de la France, iii, pp. 97. '93 GAGE, Lake and Brook Lampreys, Wilder Quarter Century Book, Ithaca, 421–493. '91 HOWES, G. B., Lamprey's affinities and relationships, P. Tr. Liverpool Biol. Soc. vi, 122–147. '89 JULIN, C., Morphologie de l'Ammocœte, B. Sci. France et Belge, 281–282. '90 KAENSCHE, C. C., Metamorphose des Ammocœtes, Schneider's Zool. Beiträge, II, 219–250.

ANATOMY, GENERAL.—'86 CUNNINGHAM, J. T., Critique of Dohrn's views of Cyclostome morphology, Q. J. M. S. xxvii, 265–284. '88 JULIN, C., Anatomie de l'Ammocœtes, B. Sci. du Nord de la France, x, 265–295. '75 LANGERHANS, P., Untersuchungen ü. Petromyzon, VH. d. n. Gesell. Friburg, XI, Heft 3. '37

MÜLLER, J., Vergleich. Anat. d. Myxinoiden, AH. K. Akad. Wiss. Berlin, 65–340, 9 pls. '79 SCHNEIDER, A., Beiträge zur vergleich. Anat. 4to, pp. 164, Berlin.

SKELETON.—'92 BURNE, R., Branchial Basket in Myxine, P. Zool. S. 706–708. '69 GEGENBAUR, C., Sketelgewebe, Jen. Z. V. '93 HASSE, C., Wirbel. Z. wiss. Zool. 290–305. '76 HUXLEY, T. H., Craniofacial Apparatus, J. Anat. Phys. x, 412–429. ('84) PARKER, W. K., Monograph of Skeleton of Petrom. and Myxine, P. Roy. S., '82, 439–443 and Phil. Tr. Roy. S., '83, 373–457. '78 PÉRÉPELKINE, K., Structure de la Notochorde, B. Mosc. liii, 107–108. '92 RETZIUS, G., Caudalskelet der Myxine, Biol. Fören. iii, 81–84.

MUSCULATURE.—'75 FÜRBRINGER, P., Muskulatur des Kopf-skelets, Jen. Z. ix, N F. II. '67 GRENACHER, H., Muskulatur, Z. wiss. Zool. xvii. '59 KEFERSTEIN (Histological), Du Bois R's. A. f. Anat. 548. '82 SCHNEIDER, A., ü. d. Rectus, Zool. Anz. N. 107, p. 164. '52 STANNIUS, H. (Heart fibres), Z. wiss. Zool. iv, 252. '52 (Histology), L'Institut, xx, 132–134, and Göttin. Nachricht. '51, 225–235. '82 STEINDACHNER, A., ü. d. Rectus. Zool. Anz. v, 660.

FINS.—'85 CLELAND, J., Tail of Myxine, R. Br. Ass. Adv. Sci. 1069.

INTEGUMENT, TEETH.—'88 BEARD, J., Teeth of Myxinoids, Nature, xxxvii, 499, and Anat. Anz. iii, 169–172. '91 BEHRENS, Hornzähne v. Myxine, Zool. Anz. xiv, 83–87. '82 BLOMFIELD, S. E., Thread. cells of Epidermis of Myx. Q. J. M. S. xxii, 355–361. '76 FOETTINGER, A., Structure de l'Epiderme, B. Acad. roy. d. Belge. ix, No. 3. '94 JACOBY, M., Hornzähne, A. mikr. Anat. 117–148. '60 KÖLLIKER, Inhalt d. Schleimsäcke d. Epidermis, Würzb. naturwiss. Zeitschr. i, 1–10. '64 MÜLLER, H., Epidermis, Würzb. naturwiss. Zeitschr. v, 43–53. '89 POGOJEFF, L., Haut, A. mikr. Anat. xxxiv, 106–122. '61 SCHULTZE, M., Kolben-förmig Gebilde, A. Anat. u. Phys.

VISCERA, GLANDS, CIRCULATORY.—'46 DUVERNOY, G. L., Sinus veineux génital, C. R. xxii, 662. '76 EWART, T. C., Abdom. pores, J. Anat. and Phys. x. '78 Vascular peribranchial spaces, J. Anat. and Phys. xii, 232–236. '89 GAGE, S. H., Blood, P. Am. S. Micros. x, 77–83, and J. R. M. S. 494. '17 HOME (Gills), Isis, 25–35. '93 HOWES, G. B., Abnormal gill clefts in Pet. and Myx. P. Zool. S. 730–733. '87 JULIN, CH. (Two anterior gill slits), B. Acad. Roy. Sci. Brux. xiii, 275–293. '88 Appareil vasculaire, Zool. Anz. xi, 567–568. '94 KERKALDY, J. W., Head-kidney of Myx. Q. J.

M. S. 353–359. '90 KLINCKOWSTRÖM, A., Darm-u. Lebervenen
b. Myx. Biol. Fören. 62–67. '93 KUPFFER, C. v., Pankreas, SB.
Morph. Gesell. Mün. ix, 37–59. '82 LEGOUIS, P. S., Pancreas,
C. R. xcv, 305–308, and Ann. S. Sci. Brux. viii, 187–304. '32
MAYER, C., milz. Gror. Nat. xxxiv, 165–166. '76 MEYER, F.,
Nieren, Centr. f. d. medicin. Wiss. No. 2. '39 MÜLLER, J.,
Gefässe : Wundernetze, Monatsheft, Berlin, 184–186, and 272–292.
'39 Lymphgefässe, AH. d. Berl. Akad. '73 MÜLLER, W.,
Urniere b. Myx. Jen. Z. vii. '75 Urogenital system, AH. d.
Ak. d. Wiss. Berl. ix. '21 OKEN, J. (Gills), Isis, 271–272, and
'29 VH. Gesell. Natur. Berl. I, 133–141. '46 ROBIN, C., Système
veineux, L'Inst. xiv, 120–123. '87 THOMPSON, D'A. W., Blood,
Ann. N. H. xx, 231–233, and Anat. Anz. ii, 630–632. '84 WELDON,
W. F., Head-kidney of Bdell. Q. J. M. S. xxiv, 171–182.

NERVOUS SYSTEM, END ORGANS.— '82 AHLBORN, F., Gött-
ing. Nachr. 677–682. '83 Z. wiss. Zool. xxxix, 191–294. '84
Hirnnerven, Z. wiss. Zool. xl, 286–308. '88–'89 BEARD, J.,
Parietal eye, Nature, xxxvi, 246–298, and 340–341, also Q. J. M. S.
xxix, 55–73. '77 FREUD, S., Nervenwurzeln im Rückenmark,
SB. Ak. Wien, lxxv. '78 Spinalganglien, SB. Ak. Wien, lxxviii.
'78 GOETTE, A., Spinalganglien, A. mikr. Anat. xv, 332–338.
'72 HASSE, Auditory organ, Anat. Studien, I, Heft 3, and Ketel,
ibid. 489–541. '79 JÉLÉNEF, A. (Cerebellum), B. Pétersb. xxv,
333–345. '86 JULIN, C. (Sympathetic), A. Zool. exper. vi, and
Anat. Anz. '87, 192–201. '87 Nerfs latéral, B. Acad. R. Brux.
xiii, 300–309. '92 KOHL, C., D. Auge v. Pet. u. Myx. Leip.
Reudnitz. br. '37 MÜLLER, J., Gehörorgan, Abh. Berl. Acad.
'38 Nervensystem d. Myx. Berl. Monatsber, 16–20. '85 NANSEN,
F., Aarsber. Berg. Mus. 55–78 (trans. in Ann. N. H. xviii,
209–226. '64 OWSJANNIKOW, P., Gehör. Mém. Acad. St. Pétersb.
viii. '83 (Sympathetic), Arb. Naturf. Gesell. xiv, and '84 B.
Acad. St. Pétersb. xxviii, 439–448. '88 (Pineal eye), Mém.
Acad. St. Pétersb. xxxvi. '86 RANSOM, W. and THOMPSON, D'A.,
Spinal and visceral nerves, Zool. Anz. ix, 421. '80 RETZIUS,
G., Riechepithel. A. f. Anat. u. Phys. '90 (Myxine : caudal
heart, nerve and sub-cutaneous ganglion cells), Biol. Untersuch.
Neue Folge, Stockholm, Fol. '92 Nervenendungen in der Haut,
des Pet., op. cit. (2) III, 37–40, and (tail structures), op. cit. iv,
36–41. '93 Gehirn-Auge v., Myx. op. cit. v, 27–30. And
Geschmacksknospen bei Pet. op. cit., 69–70. '82 ROHON, J. V.,
Ursprung d. acusticus. SB. Ak. Wien, lxxxv, 245–267. '84 SAC-
CHI, T. (Neuroglia of retina), Arch. Ital. Biol. vi, 76–96.

'80 SCHNEIDER, A., Nerven, **Zool. Anz. 330.** '71 SCHULTZE, M.,
Retina, SB. nieder-rhein Gesell. 6 Nov. '79 SELENEFF, A. (Cerebellum), B. Acad. St. Pétersb. xxv, 333–343, and '80 Mélanges
Biol. St. Pétersb. x, 307–325. '88 WHITWELL, J. R., Epiphysis,
J. Anat. Phys. xxii, 502–504. '79 WIEDERSHEIM (Brain and
nerves), Zool. Anz. 589–592 **and '80 in Jen. Z. xiv, 1–24.**

EMBRYOLOGY.

I. *General.* '88 GOETTE, A., **Zool. Anz. xi, 160–163 and '90 in AH.
3, Entwicklungesch. d. Thiere, 5 Heft. '88 KUPFFER, C. v., SB.**
bayer. Akad. I, 70–79, and '90 in A. mikr. Anat. xxxv, 469–558.
'90 NESTLER, **Zool. Anz. xiii.** '81 NUEL, J. B., A. d. Biol. i, and
'82 ii, 403–454; also J. R. M. S. ii, 26–27. '70 OWSJANNIKOW, **B.**
Acad. St. Pétersb. xiv, 325 and **'89, xxxii,** 83–95, **and '91 Mél.
Biol.** xiii, 55–67 and B. Acad. **St. Pétersb. 13–55, and '93 Ann.
N. H. xi, 30–43.** '81 SCOTT, W. B., **Q. J. M. S. xxi, 146–153, Zool.
Anz.** ('80) 422, and Morph. **JB. vii.** Also '82, A. **Zool. expér.**
ix, and (pituitary body and **teeth) Science, ii, 184–186, 731–732.**
Also '88 J. of Morph. i, 253–310. '87 SHIPLEY, A. E., **Q. J. M. S.**
xxvii, 325–370, **and** A. Zool. **expér. i.** Also Stud. Morph. Lab.
Camb. iii.

II. *Ovary, Sperm, **Fertilization.*** '93 BEARD, J., **Testes with ova,**
Anat. Anz. viii, 59–60. '87 BÖHM, A. A., Befruchtung, **SB. bayer.**
Akad. 8 Feb., and A. mikr. **Anat. xxxii.** '77 CALBERLA, E.,
Befruchtungsvorgang. **Z. wiss. Zool.** xxx, 437–486. '87 CUNNING-
HAM, T., Ova of Bdell. **Tr. Roy.** S. Edinb. xxx, 247–250, (Myxine)
Zool. Anz. 390–392, **and Q. J. M. S. xxvii,** 49–76. '91 Sper-
matogenesis of Myx. **Q. J. M. S.** xxxiii, 169–186, **and Zool. Anz.
xiv, 22–27.** '83 FERRY, L. (Internal fecundation), **CR. lxxxxvi,
721–722,** and Ann. N. H. xi, 388. '75 GULLIVER, Spermatozoa,
P. Zool. S. 336. '78 KUPFFER, C. v., u. BENECKE, **Befruchtung.
Festschr.** Th. Schwann Königsberg. '86 NANSEN (Protandric),
Myxine, Zool. Anz. 676, '87 in Bergens Mus. Aarsber, pp. 34, and
'89 (Q. J. M. S. 188–189) Zool. Anz. 261. '80 NUSSBAUM, M.
(Eggs and spawning), A. mikr. Anat. xviii, 1–121. '89 RETZIUS,
Entwick. d. Myx. Biol. For. i, 22–28, Anhang. 50–51 (Dec. '88).
'63 STEENSTRUP, J. (Egg of Myxine), Oversigt. o. d. K. danske
Videnskabernes Selskabs Forh. **233–239.** '87 WEBER, M.
(Reply to Cunningham), Zool. Anz. **318–321. Geslachtsorganen v.**
Myx. Ned. Dierkd. Vereen, **4 pp. D. I. Afl.** 3–4.

III. *Organogeny.* '77 CALBERLA, E., Medullarrohr, Morph. JB. iii.
'82 DOHRN, Hypophysis, Zool. **Anz. 587–588,** '83 in MT. z. Stat.
Neapel, iv, 172–189. '84 Visceralbogen, **op.** cit. v, 152–189.

'86 Thyreoidea, op. cit. vi. '87 (Thyroid, Spiracle, Pseudo-
branch), op. cit. vii, 301–337. '88 Nerven u. Gefässe, op. cit. viii,
233. '92 HATTA, S., Germinal Layers, J. Coll. Sci. Japan, v,
129–147. '84 HERMS, E., Nervus acusticus, SB. math. nat. Acad.
München. 333–354. '93 McCLURE, Early stages, Zool. Anz. xvi,
429. '85 SHIPLEY, A. C., Mesoblast and blastopore, P. Roy. S.
Lond. xxxix, 244–248. '86 Nervous system, P. Camb. Phil. S.
v, 374. '79 WIEDERSHEIM, Brain and spinal nerves, Zool. Anz.
ii, 589–592.

THE OSTRACODERMS AND PALÆOSPONDYLUS

(Cf. SMITH WOODWARD'S Cat. Foss. Fishes.)

'85 COPE, E. D., Position of Pterichthys, Am. Nat. xix, 289–291.
'92 CLAYPOLE, E. W., Pteraspidian family, Q. J. Geol. S. xlviii,
542–562. '90 PATTEN, W., Q. J. M. S. xxxi, 359–365. '94
Limulus and Pteraspis, Anat. Anz. '94 ROHON, J. V. (Metamerism of Cephalaspids), Zool. Anz. 51. '88 TRAQUAIR, R. H.,
Asterolepids, Ann. N. H. ii, 485–504, and ('89) P. Phys. Soc.
Edinb. 23–46. '90 Palæospondylus, Ann. N. H. vi, 485, and
'93 in P. Roy. Phys. Soc. Edinb. xii, 87–94, and '94 op. cit. xii,
312–321. '92 SMITH WOODWARD, Forerunners of Backboned
Animals, Nat. Sci. i, 596–602.

THE SHARKS

GENERAL. — '65 DUMÉRIL, A, Hist. nat. d. poissons, Tome I
(Sharks, Skates, and Chimæra), pp. 720, Paris. '79–'80 MIKLOUHO-
MACLAY, Plagiostomes of the Pacific, i, ii, iii (Anat. notes:
Dentition of young Cestracionts), P. Linn. S. N. S. Wales, iii,
306–326. '41 MÜLLER, u. HENLE, Syst. Beschr. d. Plagiost.
folio, 60 pls. Berlin.

ANATOMY, GENERAL. — '85 GARMAN, Chlamydoselache, B. Mus.
Comp. Zool. (cf. Günther in *Challenger* report). '90 JAEKEL,
O., Kiemenstellung u. System. d. Sel. (?) Berlin. '91 Pristi-
ophorus, A. f. Naturges, 15–48. LEUCKART, R., Bildung d.
Körpergestalt b. Rochen. Z. wiss. Zool. ii, 258. '93 MAREY
(Swimming movements of Ray), C. R., cxvi, 77–81. '92 MAR-
SHALL and HURST, Practical Zoölogy (Dog-fish), Putnam, N. Y.
'84 PARKER, T. J., Zoötomy (Skate), Macmillan. '84 WILS,
H. B., Squatina. Lugd. Bat.

SKELETON. — '85 DOHRN (Visceral arches of Skate), MT. z. Stat.

Neapel, vi. '72 GEGENBAUR, Kopfskelet, 4to, Leip. '78 HASSE, C. (Vertebræ, Morph. JB. iv, 214–268, and op. cit. Supplt. 43–58, and Berl. Verh. Naturw. 173–174 (Zool. Anz. i, 144–148 and 167–171). Also '85, 4to, 27 pp. Leip. '84 HASWELL, W. A., P. Linn. S. N. S. Wales, ix, 71–117. '78 HUBRECHT, Bronn's Klassen u. Ordnungen. '64 KÖLLIKER, Wirbel, Senkenb. Naturf. Gesell. v, 51–99. Also 4to. Frankfurt. '90 PARKER, T. J., Sternum in Notidanus, Nature, xliii, 142. '78 PARKER, W. K., Skull (with develop.) of shark and skate, Tr. Zool. S. x, 184–234. '84 ROSEN-BERG, E., Occiput, Festschr. Dorpat. 4to, 20 pp. And '87 in SB. Gesell. Dorp. viii, 31–34. '73 STEENSTRUP, J. (Gill rakers of Selache), Overs. Dan. Selsk. No. 1. '90 WHITE, P. J., Skull and visc. skelet. of Læmargus, Anat. Anz. v, 259–261.

FINS AND GIRDLES. — '81 BALFOUR, Skates, P. Zool. S. '70 GEGENBAUR, Skel. of, Jen. Z. v, 397–447, also (claspers) 448–458. '90 HOWES, Batoid and Squaloraja, P. Zool. S. 675–688. '85 MAYER, P., Unpaaren Flos. MT. z. Stat. Neap. vi, 217–285. '79 METSCHNIKOFF, O. (Morph. of girdles), Z. wiss. Zool. xxviii, 423–438. '79 MIVART, ST. G., T. Zool. S. x, 439–484. '92 MOLLIER, S. (Entwickelung), Vorl. MT. Anat. Anz. vii, 351–365, also '77 THACHER, J., Tr. Connec. Acad. iii.

INTEGUMENT AND TEETH. — '44 AGASSIZ, L., Dents et rayons, 4to, Neuchâtel. '68 HANNOVER, A., Écailles. (?). HARLESS, E., Zahnbau v. Myliobates, 4to (?). '74 HERTWIG, O. (also development), Jen. Z. viii. '72 RANVIER, L., Étranglements annulaire (Rays), C. R. 1129–1132.

VISCERA, VESSELS, GLANDS. — '92 ANTIPA, G., Thymus, Anat. Anz. vii, 690–692. '88 AYRES, H., Carotids (Chlamydosel.), B. Mus. Comp. Zool. xvii, 191–224. '78 BLANCHARD, R., Superanal gland, J. de l'Anat. Phys. xiv, 442–450, and B. S. Zool. Fr. vii, 399–401. '79 BRIDGE, Pori abdominales, J. Anat. Phys. '82 DRÖSCHER, W. (Histol. of gills), A. f. Nat. xlviii, 120–177. '90 EWART, J. C., Spiracle of Lamna, J. Anat. Phys. xxiv, 227–229. '78 FÜRBRINGER, Excretory system, Morph. JB, 49–56. '91 GEGENBAUR, Cöcalanhänge v. Mitteldarm, Jen. Z. (?), 180–184. '90 HOWES, Kidney of Raja, J. Anat. Phys. xxiv, 407–422. '90 Visceral anat. of Hypnos, P. Zool. S. 669–675. '72 HYRTL, J., Kopfarterien, DS. Ak. Wiss. Wien, xxxii, 263–275. '85 LIST, J. H., Cloakenepithel, SB. Akad. Wien (?), and Anat. Anz. vii, 545–546. '80 PARKER, T. J., Spiral valve of Ray, T. Zool. S. xi, 49–61, and Venous system, Tr. N. Zeal. Ins. xiii, 413–418. '86 On the vessels of Mustelus, and ('87) on Carcharodon, P. Zool.

S. (?). '90 PILLIET, A. (Histol. of liver), C. R. Soc. Biol. ii, 690–694. '82 POUCHET, G., Termin. vascul. d. la Rate, J. de l'Anat. Phys. xviii, 498–502. '88 RÜCKERT, J., Excretions-organe, A. Anat. Phys. (Anat.), 205–278. '77 STOHR (Valves in arterial conus), Morph. JB. ii, 197–228. '79 TROIS, E. F. (Carotids of Oxyrhina), Atti del Inst. Ven. and '83 (Of Alopecias), op. cit. '73 TURNER, W., Visceral Anat. of Læmargus, J. Anat. Phys. 233. '75 Spiny Shark, op. cit. '79 Pori abdominales, op. cit. xiv, 101–102. '81 Teeth and gill-rakers of Selache, op. cit. xiv, 273–286. '93 VIRCHOW, H., Spritzlochkieme, SB. Gesell. n. Freunde, Berl. 177–182; Augengefässe ('90) A. Anat. Phys. (Phys.), 169–173. '67 MIKLUCHO-MACLAY, Schwimmblasenrudiment, Jen. Z. iii, 448–453.

NERVOUS SYSTEM AND END ORGANS. — '92 BRAUS, H., Rami ventrales d. vord. Spinalnerven. Inaug. Diss. 35 pp. Jena. '83 CATTIE, J. F., Epiphysis, Z. wiss. Zoöl. xxxix, 720–722, and A. Biol. iii, 101–196. '90 CHEVREL, R., Sympathetic, A. Zool. expér. v, 196 pp. '91 COGGI, A., Vésicules de Savi (Torpedo), A. Ital. Biol. xvi, 216–224. '92 EDINGER, L., Zwischenhirn, AH. Senck. Gesell. xviii, 55 pp. (and Anat. Anz. vii, 472–476). '78 EHLERS, E., Gehirn u. Epiphyse, Z. wiss. Zool. xxx, Suppl. 607. '90 EWART, J. C., Cran. nerves of Torpedo, P. Roy. S. xlvii, 240–291. '92 Electric organ of Skate, Phil. Trans. clxxxiii, 389–420 (Abs. in P. Roy. S. 474–476). '92 Sensory canals of Læmargus, and Skate, Tr. Roy. S. Edinb. Nos. 5–6 (Abs. in Zool. Anz. xv), 116–120. '88 FRITSCH, G., Bedeuting d. Kanalsyst. SB. Ak. Berl. 273–306. '88–'89 GARMAN, Lateral line (figures many forms), B. Mus. Comp. Zool. xvii, No. 2. '71 GEGENBAUR, Kopfnerven v. Hexanchus, Jen. Z. 497–560. '75 JACKSON, W. H., and CLARKE, Brain and nerves, J. Anat. Phys. x, 75–197. '87 LÉGER, M., Cervelet d'un Alopias, B. S. Philom. xi, 160–163. '81 MARSHALL, A. M., Head cavities and nerves, Q. J. M. S. xxi, 72, and (with SPENCER) 469–499. '80 MERKEL, Sens. nerven in d. Haut. Rostock. '70 MIKLUCHO-MACLAY, Gehirn, 4to, 74 pp. Leip. '78 RETZIUS, G., Memb. Gehörlabyr. A. Anat. Phys. (Anat.), 83–105. '91 REX, H. (Morph. of cran. nerves), Morph. JB. xvii, 417–466. '78 ROHON, J. (Brain), DS. Ak. Wien, xxxviii, 43–108. '86 SANDERS, A., Phil. Trans. clxxvii, 733–764, and P. R. S. xl. '89 SHORE, T. W., Vagus, J. Anat. Phys. iii, 428–451. '80 SOLGER (Organs of lateral line), A. mikr. Anat. xvii. '73 STIEDA, L., Rückenmark, Z. wiss. Zool. xxiii, 435–442. '73 TODARO, F. (Gustatory buds and branch. membr.), Récherche lab. univers. Roma. (Abs. in A. Zool. expér.) 534-558. '91 VALENTI, G., Histogenesis, A. Ital. Biol. xvi, 247-252. '76 VIAULT, F., Histology, A. Zool. expér. v, 441-528. '83 VIGNAL, Sys. ganglionaire (Ray), op. cit. i, xvii.

EMBRYOLOGY, GENERAL. — '76 BALFOUR, G. M., in Q. J. M. S. xiv, and Camb. J. Anat. Phys. xi, pt. I. Also '78 Monograph, Macmillan, London. '90 BEARD, J., Skate, R. Fish. Board, Edin. '89 EIGENMANN, C. H. and R. S., Young stages, West Am. Nat. vi, 150-151. '85 HASWELL, W. A., Young Pristiophorus, P. Linn. S. N. S. Wales, ix, 680-681. '77 HIS, W., Ueb. d. Bildung v. Haifishembryonen, Z. wiss. Zool. ii, 108-124. '81 HOFFMANN, C. K. (?), Harlem. '88 KASTSCHENKO, N., Anat. Anz. iii, 445-467. '70 KOWALEWSKY, A. (Russian), Trans. Kiew. Soc. of Nat. i. '52 LEYDIG, F., Mikr. Rochen u. Haie, Leip. Englemann, iv, 127 pp. '76 MALM, A. W., Kongl. vet. akad. förhand. Stockholm. '42 MÜLLER, J. (Emb. differences of Shark and Ray), fol. Berlin. '86 PERENYI, J., Torpedo, Zool. Anz. ix, No. 227, 433-436. '77 SCHULTZ, A., A. mikr. Anat. '92 SEDGWICK, A., Q. J. M. S. xxxiii, 559-586. '64, WYMAN, Raja, Mem. Am. Acad. Arts and Sciences, ix.

I. *Breeding, Gestation.* — '73 AGASSIZ, A., P. Bost. Soc. xiv, 339. '90 ALCOCK, A., J. A. Soc. Bombay, lix, 51-56. (Of Rays) '92 Ann. N. H. ix, 417-427, and x, 1-8. '79 BOLAU, H. (Of Scyllium), VH. Ver. Hamb. iii, 122-130 and ('81) Z. wiss. Zool. xxxv. '67 COSTE (Of Scyllium), C. R. 99-100, and Ann. M. N. H. xix, 227. '70 HOME, E., Ovoviviparous Sharks.—— '68 MACDONALD, J. D., and BARRON, CH., Heptanchus, P. Zool. S. 371-373 '85 MATTHEWS, J. D., Oviduct of Skate, J. Anat. Phys. xix, 144-149. '90 MEHRDORF, C. (Gestation), Rostock, 51 pp. '72 MEYER (Scyllium), Zool. Gart. 371. '85 OERLEY, L. (Gestation), Term. füzetek. ix, 293-309. '90 PARKER, T. J., Mustelus, Tr. N. Zeal. Inst. xxii, 331-333. '77 PETRI, K. R., Copulationsorgane, Z. wiss. Zool. xxx, 288-385. '75 SCHENK, S. L., Kiemenfäder, SB. Akad. Wien, 12 pp. '79 SCHMIDTLEIN, R., MT. z. Stat. Neap. l, 61. '91 WOOD-MASON and ALCOCK, Gestation of Pteroplatæa, P. Roy. S. xlix, 359-367, and l, 202-209. '67 TROIS, E. F. (Acanthias' gestation), Atti. Inst. Ven. 171-176. '77 (Gestation of Myliobatis and Centrina), op. cit. ii, 429. '77 TURNER, W., Læmargus' oviducts, J. Anat. Phys. xii, 604-607, and ('85) op. cit. xix, 221-222.

II. *Egg, Gastrulation.* — '92 DOHRN, Schwann'schen Kerne, Anat. Anz. vii, 348-351. '72 GERBE, Z., Segmentation, J. de l'Anat.

R

Phys. '81 HERRMANN, G., Spermatogenèse, C. R. xciii, 858–860,
and '83 J. l'Anat. Phys. xviii, 373–432. '83 HERTWIG, O.
(Middle germ layer), Jen. Z. xvi, 287–290. '83 HOFFMANN, C. K.
(Middle germ layer), Arch. Neerland. xviii, 241. '92 Endothelial
Anlage d. Herzens, Anat. Anz. vii, 270–273, and '93 in Morph.
JB. xix, 592–648. '88 KASTSCHENKO, Dotterkerne, Anat. Anz.
'90 (Early develop. and muscles), Zool. Beitr. ii, 251–266. '86
KOLLMANN, Fürchung (C. R. '84), Congr. pér. internat. d. sc.
méd. Copenhague, i, Sect. d'Anat. 50–52, and VH. d. Natur.
Gesell. Basle, Th. viii, Heft 1. '78 LA VALLETTE, ST. GEORGE, A.,
Spermatosomatum, Bonn, 4to, 9 pp. '79 LUTKEN, Læmargus'
eggs, oviduct, Vid. Medd. ('80) 56–61. '84 PERRAVEX, M. E.
(Egg case), C. R. xcix, 1080–1082. '89 OSTROUMOFF, A., Blasto-
porus u. Schwanzdarm, Zool. Anz. xii, 364–366. '85 RÜCKERT, J.,
Keimblattbildung, SB. Gesell. Morph. München, i, 48–104, and
'89 in Anat. Anz. iv, 353–374. '86 Gastrulation (and middle
germ layer), Anat. Anz. 286–287, and '87 op. cit. 97–112 and
151–174. '91 Befruchtung, Anat. Anz. vi, 308–322. '92 Ova-
rialei, vii, 107–158, Chromosomen, viii, 44–52. '86 RYDER,
Segmentation, Am. Nat. xx, 470–473, and B. U. S. F C. 8–10.
'82 SABATIER, A., Spermatogenèse, C. R. xciv, 1097–1099. '75
SCHULTZ, A. (Ovogenesis), A. mikr. Anat. xi, 569–582. '73
SCHENK, S. L. (Egg and oviduct), SB. Akad. Wien, lxxiii.
74 Dotterstrang, op. cit. lxix, 301–308. '90 SCHNEIDER, A.
(Gastrula-muscles), Zool. Beitr. ii, 251–266. '85 SWAEN, A.
(Germ layers and blood), B. Acad. roy. Belgique, ix, and '86 in
A. de Biol. vii, 537–585. '83 TROIS, E. F., Spermatozoi, Atti. Inst.
Ven. and J. Microgr. vii, 193–196. '84 VAILLANT, L., Orientation
des œufs dans l'utérus, B. Soc. Philom. viii, 178–179. '88 ZIEGLER
(Mesenchyme), A. mikr. Anat. xxxii. '92 ZIEGLER, H. E. and F.
(Early development), A. mikr. Anat. xxxix, 56–102.

III. *Integument, Skeleton.* — '81 BENDA, C., Dentinbildung, A. mikr.
Anat. xx, 246–270. '79 HASSE, C., Knorpel, Zool. Anz. ii,
325–329, 351–355, and 371–374. '82 Wirbelsäule, Jena ('79),
4to. '92 Wirbelsäule, Z. wiss. Zool. lv, 519–531. '60 KÖLLIKER,
A., Chorda u. Wirbel. Würz. '87 PERENYI, J., Chorda, peri-
chordal. Math. u. Naturwiss. Ber. a. Ungarn, iv, 214–217, and
('89) in 218–241. '93 PLATT, JULIA B., Ectodermic cartilage,
Anat. Anz. 506. '78 REICHERT, Vordere Ende d. Chorda, AH.
Ak. Berl. 49–113. '84 ROSENBERG, E., Occipitalregion, Festschrift,
Dorpat, 26 pp. 4to.

IV. *Viscera.* — '87 BEARD, J., Segmental duct, Anat. Anz. ii, 646–652.

'85 BEMMELEN, J. F. v. (Rudimentary gill slits), MT. z. Stat. Neap. vi, 165–184. '79 BLANCHARD, R., Fingerförmigen Drüse, MT. Emb. Inst. Schenk, iii, 179–192, and ('77) J. de l'Anat. xlv, 442–450. '84 DOHRN, Kiemenbogen, Flossen, MT. z. Stat. Neap. v, 102–189. '87 MAYER, P (Circulatory), op. cit. vii, 338–370, and ('88) viii, 307–373. Also Anat. Anz. ix, 185–192. '77 MAYER, F., Urogenitalsys. SB. Gesell. Leip. ('76), 38–44. '92 RABL, C., Venensys. Leuckart Festschr. 228–235. '92 RAF-FAELE, F., Sist. vascolare, MT. z. Stat. Neap. x, 441–479. '88 RÜCKERT, J., Endothel. Anlagen d. Herzens, Biol. Centralbl. viii. '89 Excretionssys. Zool. Anz. xii, 15–22. '75 SEMPER, C., Urogenitalsys. Arb. a. d. Zool. Zoot. Inst. Würz. ii. '88 WIJHE, J. W. v., Excretionsorgane, Zool. Anz. xi, 539–540, and Anat. Anz. iii, 74–76, and '89 in A. mikr. Anat. xxxiii, 461–516.

V. *Nervous System and End Organs.* — '85 BEARD, J., Cranial ganglia, Zool. Anz. viii, 220–223, Anat. Anz. iii, 874–905, and op. cit. '92, 191–206. '91 KILLIAN, Metamerie, VH. Anat. Gesell. 85–107. '88 DOHRN (Motor fibres), MT. z. Stat. Neap. viii, 441–462. '91 Augenmuskelnerven, op. cit. x, 1–40. '91 FRORIEP, Kopfnerven, VH. Anat. Gesell. 55–65. '92 LENHOSSÉK, M. v. (Spinal ganglia and cord), Anat. Anz. vii, 519–539. '85 ONODI, A. (Nerve roots), Ber. Math. Nat. Ungarn, ii, 310–336. '89 OSTROUMOFF, A., Froriep'schen Ganglien, Zool. Anz. xii, 363–364. '90 PLATT, J. B., Anterior head cavities, Zool. Anz. xiii, 239, and '91 J. of Morph. v, 79–106, and Anat. Anz. vi, 251–265. '96 PUNIS, G. C., Pineal eye, P. Phys. Soc. Edinb. 62–67. '92 RABL, C., Metamerie, VH. Anat. Gesell. 104–135. '80 RABL-RÜCKHARDT (Metamerism), Morph. JB. vi, 535–570. '93 Lobus olf. impar. Anat. Anz. Sep. 15. '83 VIGNAL, W., Système gang-lionaire, A. Zool. expér. i, 17–20. '83 VAN WIJHE (Metamerism), VH. Akad. Wiss. Amsterdam. '76 WILDER, B. G., Anterior brain mass, Am. J. Sci. xii, 103–106.

MORPHOLOGY OF FOSSIL SHARKS. — V. ref. in S. Wood-ward's Catalogue, also in present writer's article on Cladoselache, '94 J. of Morph. ix, 112. In addition, '88 BROGNIART ET SAUVAGE, Études sur le Terrain Houiller de Commentry, Liv. iii, B. de la S. d. l'Indus. minér. ii, 1–39. '93 CLAYPOLE, E. W., Cladodonts, Am. Geol. 325–331, and ('95) op. cit. Jan. '93 COPE, Clado-donts, Am. Nat. Sept. Also '94 J. Am. N. S. Phila. ix, 427–441. '92–'94 DAVIS, J. W., Pleuracanths, Acanthodians, Tr. Dub. Roy. S. '94 DEAN, B., Cladodont, Tr. N. Y. Acad. Sci. 115–119. '92 JAEKEL, O. (Eocene sharks), SB. Gesell. Nat. Fr. Berlin, p. 61, and Cladodus, l.c. 156–158. **'95** SMITH WOODWARD, Primæval sharks, Nat. Sci. vi, 38–44.

THE CHIMÆROIDS

(Cf. esp. DUMÉRIL, Ref. p. 238.)

'94 BEAN, T. H., Harriotta, P. U. S. Nat. Mus. xvii, 471–473. **'52** COSTA (Anatomy), Faun. regno Napoli. **'51** LEYDIG, F., Anat. and Hist. Mül. A. f. Anat. Phys. xviii, 241–271. **'76** HUBRECHT, A., Kopfskelet, Nied. A. Zool. iii, 255–276, and **'77** in Morph. JB. iii, 280–282. **'86** PARKER, T. J., Claspers of Callorhynchus, Nat. xxxix, 635. **'75** SOLGER, B. (Visceral skeleton), Morph. JB. I, H. I. **'37** DUVERNOY, G. Z. (Heart and vessels), Ann. d. Sc. Nat. 1–16. **'78** LANKESTER, E. R., Heart, P. Zool. S. 634, and **'79** in Tr. Zool. S. x, 493–506. **'42** MÜLLER, J. (Nerves and heart, critique of Valentin), A. f. Anat. ('43) ccliii. **'89** GARMAN, S., Lateral line, Mus. Comp. Zool. xvii. **'70** MIKLUCHO-MACLAY (Brain), Jen. Z. v, 132. **'79** SOLGER, B. (Lateral line), A. mikr. Anat. xvii, 95–113. **'42** VALENTIN (Brain and Nebenherzen), A. mikr. Anat. 25–45. **'77** WILDER, Brain, P. Philadel. Acad. Sci. 219–250. **'90** ALCOCK, Egg capsule of Callorhynchus, Ann. N. H. viii, 22. **'71** CUNNINGHAM, Callorhynchus' egg, Notes on the N. H. of the Straits of Magellan, 340. **'89** GÜNTHER, Chimæra's egg. A. N. H. iv, 275–280.

For literature of Fossil Chimæroids v. SMITH WOODWARD'S Catalogue.

THE LUNG-FISHES

GENERAL (NATURAL HISTORY). — **'94** BOHLS, Fang u. Lebensweise v. Lepidosiren. Nachr. Gesell. Göttingen, 80–83. **'76** CASTLENAU, F., Ceratodus, C. R. lxxxiii, 1034. **'92** DUBOIS, R., Respiration, "hibernation," Ann. S. Linn. Lyon, xxxix, 65–72. **'66** DUMÉRIL, A. M. C., C. R. 97–100, and Ann. N. H. xvii, 160. **'70** (Swim-bladder, etc.), Angers ? **'94** EHLERS, E., Lepid, n. s. Nachr. Gesell. Göttingen. FRITSCH, A. (Living and Fossil Lung-fishes and their affinities), Prag. 4to. **'87** GIGLIOLI (Rediscovery of Lepidosiren), Nat. xxxv, 343, and **'88**, Nat. xxxviii, 112. **'56** GRAY, J. E., "Lepidosiren," P. Zool. S. Lon. 342. **'88** HOWES, Rediscovery of Lep. Nat. xxxviii, 126. **'41** JARDINE, W., Ann. N. H. vii, 24. **'64** KRAUSS, F., Protopterus, Würt. n't'rwiss. Jahresber. 126–133. **'70** KREFFT, Ceratodus, P. Zool. S. 221–

224, **and** Ann. N. H. 221–224, **and** ('71) **P. Roy. S. 377.**
'91 LACHMAN, H., Protop. **Zool.** Gart. xxxii, 129. **'73 MARNO,**
E., Protop. Zool. Gart. 44. **'58–'59** McDONNELL, R., **Protop. Z.**
wiss. Zool. x. **'37** NATTERER, J., **Lepid.** Ann. Wien. **Mus. II.**
'94 NATURAL SCIENCE, **Lepid.** 324–325. **'39–'41** OWEN, **Lepi-**
dosiren annectans, Tr. Linn. **S. xviii.** '45 PETERS, W., **Protop.**
Mül. A. **'76** RAMSEY, E. P., Cerat. P. Zool. S. 698. SCHMELTZ,
Cerat. J. Mus. Godeffr. viii, 138. '66 SCLATER and BATES, Lepid.
P. Zool. S. 34. '92 SPENCER, W. B., Cerat. Vict. Natural. Melb.
June 10, and P. Roy. S. Vict. iv, 81–84. **'89** STUHLMAN, F.,
Cerat. SB. Akad. Wiss. Berl. 32. **'87** WIEDERSHEIM, **Protop.**
Anat. Anz. ii, 707–713, and R. Br. Ass. 738–740.

ANATOMY, GENERAL. — '85 AYERS, H., **Jen.** Z. Naturwiss. **xviii,**
479–527. **'87** BAUR, G., Lepid. Zool. **JB. ii, 575.** '40 BISCHOFF,
T., Lepid. **Leip.** '71 GÜNTHER, **Ceratodus, Ann. N.** H. vii, **227**
and Phil. **Trans. ('72) clxi, 511–571, and P. Roy. S. 377–379,**
Nat. Nos. 99, **100, 102.** '76 HUXLEY, **Ceratodus, P. Zool. S.**
24–58. '64 KLEIN, Protop. **Würt. n't'rwiss. Jahresber. 134–144.**
'78 MIALL, L., Cerat. and Protop. Palæont. **S. xxxii, 1–32. '88**
PARKER, W. N., Ber. d. Naturforsch. **Gesell. Friburg,** VB. **iv, H.**
3, Nat. xxxix, 9–21, **and** Tr. Cardiff Nat. **S. xx.** '91 Protop.
P. Roy. S. xlix, **549–554.** '92 Protop. (Large memoir), Tr. R.
Irish Acad. xxx, **115–227. '66** PETERS, Monatsber. **Ak.** Wiss.
Berl. 12–13.

SKELETON. — '93 KLAATSCH, H., **Wirbel, VH.** Anat. Gesell, **130–**
132. '91 TELLER, F., Skull of **Ceratodus,** AH. Geol. **Reichanst.**
xv, H. 3.

MUSCLES. — '72 HUMPHREY, G. M., **Ceratodus and Protop. J. Anat.**
and Phys. vi.

FINS AND GIRDLES. — '86 ALBRECHT, P., Protop. Fin forked, **SB.**
Ak. Berl. 545–546. '91 BOULENGER, Protop. Renewed pectoral.
'83 DAVIDOFF, M., Cerat. Pelvic fin. '84 GILL, T., Shoulder
girdle, Ann. N. H. xi, 173–178. '83 HASWELL, W. A., Cerat.
Paired fins, P. Linn. S. N. S. **Wales, vii, 2–11.** '91 HOPLEY, C.,
Protop. **Renewed** pectoral, Am. Nat. xxv, 487. **'87** HOWES, G.
B., Cerat. Paired fins compared **with sharks', P. Zool. S. 3.**
'94 LANKESTER, E. Ray, Lepid. Villous processes **of** hind limbs,
Nat. Apr. 12. **'86** SCHNEIDER, A., Zool. Anz. **ix, 521–524,** and
('87) Zool. **Beitr. ii, 97–105.** '71 TRAQUAIR, R. H., Protop. **Tail**
restored, **Br. Ass. R. '90** VANHÖFFEN, **Cerat. VH.** Gesell. **D.**
Naturf. ii, **134.**

INTEGUMENT AND TEETH. — '87 BÖCKLEN, H., Cerat. Dentition, JH. Ver. Würt. xliii, 76-81. '60-'61 KÖLLIKER, A., Protop. Histol. of Skin, Z. Naturwiss. Würzb. i. '65 PAULSON, M., Protop. Histol. of epidermis, B. Acad. Sci. St. Pétersb. viii, 141-145. '92 RÖSE, C., Zahnbau u. Zahnwechsel, Anat. Anz. vii, 821-839. '89 WALTHER, G., Prot. Skin, Z. f. Phys. Chem. xiii, H. 5. '80 WIEDERSHEIM, R., Scales, A. mikr. Anat. xviii.

VISCERA, VESSELS, GLANDS. — '80 BOAS, E. V. (Heart and arteries), Morph. JB. vii, 321-354. '78 FURBRINGER (Excretory), Morph. JB. iv, 60. '76 HUXLEY, Anterior nares, P. Zool. S. 180. '45 HYRTL, J., Lepid. AH. d. böhm Gesell. Prag. '78 LANKESTER, E. R., Heart, P. Zool. S. 634, and ('79) Tr. Zool. S. x, 493-506. '89 PARKER, W. N., Veins (L. cardinal), P. Zool. S. 145-151. '94 SPENCER, W. B., Cerat. Vessels (complete memoir), Macleay Mem. Vol. Linn. S. N. S. Wales, 2-32.

NERVOUS SYSTEM, END ORGANS. — '82 BEAUREGARD, H. (Cranial), J. de l'Anat. Phys. xvii, 230-242. '91 BURCKHARDT, R., Zirbel. Anat. Anz. vi, 348-349. '92 Cent. nerv. sys. Berlin, 64 pp. Also Zool. Gesell. ii, 92-95, and SB. Nat. Fr. Berl. 23-25. '94 (Zwischenhirndach), Anat. Anz. 152. '86 FULLIQUET, G. (Brain), A. Sci. Naturelles, xv, 94-96, and Rec. Zool. Suisse, iii, 1-130. '94 PINKUS, F. (Undescribed nerve), Anat. Anz. ix, 562-566, and (Cranial nerves of Protop.) Morph. Arb. (Schwalbe), 275-346. '89 SANDERS, A., Cent. nerv. sys. Cerat. Ann. N. H. iii, 157-188. '80 WIEDERSHEIM, Skel. and cent. nerv. sys. Jen. Z. xiv, and Morph. Stud. Heft 1, Jena. '82 WIJHE, J. W. van, Visceralskel. u. d. Nerven. Cerat. Nied. A. Zool. v, 207-320. '87 WILDER, B., Brain, Am. Nat. xxi, 544-548.

EMBRYOLOGY. — '86 BEDDARD, F. E., Ovarian ovum, P. Zool. S. 272-292, and Zool. Anz. ix, 635-637. '84 CALDWELL, W. H., (Preliminary), J. and P. Roy. S. N. S. W. xviii, and ('87) in Phil. Trans. clxxviii. '93 HASSE, C., Wirbelsäule, Z. wiss. Zool. lv, 533-542. '93 SEMON, R. (Habits and development — surface views of eggs and larvæ), DS. d. Med. Nat. Gesell. Z. Jena, pp. 50.

THE GANOIDS

GENERAL (NATURAL HISTORY). — '70 DUMÉRIL, Aug. Tome ii, pp. 625, Roret, Paris. '35 HECKEL, J., Scaphirhynchus, SB. Akad. Wien. '71 LÜTKEN (Classification), Transl. in Ann. N. H. 329-339. '85 ORR, H. (Phylogeny) Inaug. Dissert. Jena, 37 pp. '65-'66 SMITH, J. A., Calamoichthys, P. Roy. S. Edinb. v, 654-

659, and ('66) 457-479. '69 STEINDACHNER, F., Polypterus, SB. Wien. Akad. lx.

GENERAL ANATOMY.—'87 TWANZOW, N., Scaphirhynchus, B. S. Mosc. 1-41. '54 LEYDIG (Histology of Polypterus), Z. wiss. Zool. v. '50 LITTANY, M., Acipenser, B. S. Mosc. xxiii, 389-445. '44 MÜLLER, J., Bau u. Grenzen, A. f. Anat. and ('46) AH. d. Berl. Akad. d. Wiss. '92 POLLARD, H. B., Polypterus, Anat. and phylogeny, Morph. Jb. V, 387-428, and preliminary in ('91) Anat. Anz. vi, 338-343. '48 WAGNER, A., de Spatulariarum Anat. Inaug. Diss. Berol. '75 WILDER, B., Notes on Am. Gan. I. Respir. of Lepid. and Amia. II. Tail formation of Lepid. III. Pect. fin formation of Lepid. IV. Brains of Amia, Lepid., Acip., and Polyod. P. Am. Ass. Adv. Sci. xxiv, 151-193. '76 Brains, Philadel. Acad. P. xxxviii, 51-53. '78 Amia and Lepid. rudimentary spiracle, P. Am. Ass. (unpub'd), and Am. Nat. xix, 192. And in the respiration of Amia, P. Am. Ass. 306-313. '85 WRIGHT, R. R., Notes on anat. of fishes: A. Cutansense organs. B. Spiracular cleft of Amia and Lepid. C. Aud. organ of Hypophthalmus, Am. Nat. xix, 187-190 and 513. D. Hyomand. clefts and pseudobranchs of Amia and Lepid. and Amia, J. Anat. Phys. xix, 477-497. E. Amia's serrated appendages, Sci. iv, 511. '87 ZOGRAFF, N., Monograph (Russians) on Sturgeon, Tr. S. Nat. Mosc. lii, pp. 72. '87 Affinities of ganoids, Nat. xxxvii, 70.

SKELETON.—'77 BRIDGE, T., Cranium Amia J. Anat. Phys. xi, 605-622. '78 Polyodon, Phil. Trans. clxix, 683-734. '89 Cranial anat. Polypterus, P. Birmingham, Phil. S. vi, 118-130. '83 CAPAUEK, F. (Prag.). '47 FRANQUE, H., Amia, Folio, Berolini. '78 GOETTE, A., Wirbelsäule, A. mikr. Anat. x, 442-641. '93 HASSE, C., Wirbelsäule, Z. Wiss. Zool. 76-90. '60 KÖLLIKER, Ende d. Wirbelsäule, Leip. '20 KUHL u. HASSELT Ost. of Sturgeon, Kuhl's Beitr. Zool. in Vergl. Anat. 2 Abth. 188-202. '51 MOLIN, R., Scheletro dell. Acipenser, SB. Acad. Wien, vii, 357-378. '82 PARKER, W. K., Skull (and develop.) of Acip. P. Roy S. 142-145, of Lepidos. 443-491. '83 SAGEMEHL, M. (Skull of Amia), Morph. JB. ix, 177-227. '92 SCHMIDT, L. (Vertebræ of Amia), Z. wiss. Zool. liv, 748-764. '85 SHUFELDT, R. W., Amia, R. U. S. F. C. ('83) 747-834. '70 TRAQUAIR, R. H., Calamoichthys, J. Dub. Geol. Soc. June 8. '70 Skull of Polypterus, J. Anat. Phys. v, 166-183. '82 WIJHE, J. W. VAN, Visceralskelet (u. Nerven)—includes Ceratodus,—Nied. A. Zool. v, 207-320.

MUSCLES.—'85 McMurrich, J. P., Head of Amia, Stud. Biol. Lab.
J. Hop. Univ. iii, 121-153. '82 Schneider, H., Augenmuskeln,
Jen. Z. xv, 215-242.

INTEGUMENT, TEETH.—'78 Barkas, W., Teeth of Lepid. Tr.
Roy. S. N. S. Wales, xi, 203-207. '77 Mackintosh, H. W.,
Scale of Amia. (?). '80 Pawlow, H., Teeth of Sturgeon, Arb.
St. Pétersb. Nat. Gesell. No. 9, 494-508. '59 Reissner, Schup-
pen v. Polyp. and Lepid. A. f. Anat. '87 Zograff, N., Zähne d,
Knorp. gan. Biol. Centralb. vii, 178-183 and 224.

VISCERA.—'86 Cattaneo, G., Glandula gastriche nell' Acip. Rend.
Inst. Lomb. xix, 676-682. '78 Fürbringer, Excretory sys. Morph.
JB. iv, 56-60. '72 Hertwig, R., Lymph. Drüsen d. Störherzens,
A. mikr. Anat. ix, 62-79. Hoeven, J. v. D. (Air-bladder of
Lepid.) 4to. (?). '91 Hopkins, G. S., Structure of stomach
of Amia, P. Am. Micr. S. xxii, 165-169. '92 Diges. tracts of
N. A. Gan. P. Am. Ass. xli, 197. '69 Hyrtl, J., Blutgefässe
d. aus. Kiemendeckel-Kieme v. Polyp. SB. Ak. Wien, lx, 109-113.
'86 Macallum, A. B., Diges. tract and pancreas of Acip., Amia,
Lepid. J. Anat. Phys. xx, 604-636. '91 Semon, R., Zusammen-
hang d. Harn- und Geschlechtsorgane, Morph. JB. xvii, 623-635.
'77 Stöhr (Valves in conus—compares sharks'), Morph. JB. ii,
197-228. '90 Virchow, H., Spritzlochkieme v. Acip. A. Anat.
Phys. (Phys. Abt.) 586-588. '86 Wilder, Serrated appendages
of Amia, P. Am. Ass. xxxiv, 313-315.

FINS.—'94 Gegenbaur, Flossenskelet d. Crossopterygier, Morph.
JB. xxi, 119-160. '82 Rautenfeld, E. V., Skel. hint. Gliedmassen, Inaug. Diss. Dorpat, 47 pp. '77 Thacher, J., Ventral
fins, Tr. Connec. Acad. iv, 233-242. '80 Davidoff, M. V.,
Skel. d. hint. Gliedmassen, Morph. JB. vi, 126-128 and 433-468.
'66 Huxley, Illus. of struc. of Crossopt. 4to, Lon.

NERVOUS SYSTEM, END ORGANS.—'89 Allis, Lateral line
of Amia, J. of Morph. '83 Cattie, J. T., Epiphysis, Z. wiss.
Zool. xxxix, 720-722, and A. de Biol. iii, 101-196. '94 Collinge, W. E., Sensory canals, of Polypterus, P. Birmingh. S.
viii, 255-262; of Lepid. op. cit. 263-272; of Polyodon Q. J. M. S.
xxxvi, 499-437. '83 Dogiel, A., Retina, A. mikr. Anat. xxii, 419-
472, and '84 Naturf. Ges. Kasan, xi, 124 pp. '86 Geruchsorgan, Tr.
Kasan. Univ. xvi, 82 pp. '88 Retina, Anat. Anz. iii, 133-143.
'79 Gisow, A., Gehörorgan, Bonn. '81 Gisow, A., Gehörorgan,
A. mikr. Anat. xviii, 484-519. '88 Goronowitsch, N., Gehirn
u. Cranialnerven v. Acip. Morph. JB. xiii, 427-514. '70

MIKLUCHO-MACLAY, **N. v.**, Mittelhirn, Leip. **4to, pp. 74. '81**
RETZIUS, Gehörorgan, v. Polyp. Stockh. '81 SCHNEIDER, H.,
Augenmuskelnerven, Jen. **Z.** viii, 215–242. **'87** WALDSCHMIDT,
J., Centr. nerv. u. Geruchsorg. **v. Polyp. Anat. Anz. ii, 308–322.**

DEVELOPMENT. — **'78** AGASSIZ, A. (Larvæ **of Lepid.), P. Am.**
Acad. A. and Sc. xiii, 65–76. '89 ALLIS, E. P., Lateral **line,**
Amia, J. of Morph. '81 BALFOUR and PARKER, Str. **and devel. of**
Lepid. P. Roy. S. xxiii, 112–119, and **'82** in Phil. Trans. **(large**
memoir) '89 BEARD, J., Early devel. of Lepid. P. R. **S. xlvi,**
108–118. '95 DEAN, B., Early devel. gar and sturgeon, J.
Morph. xi, No. I, 1–62. '82 DUNBAR, G., Breeding **of** Lepid.
Am. Nat. May. '94 FÜLLEBORN, F. (Breed. habits Amia and
Leipd.), SB. Akad. Wiss. Berl. xl, 1–14. **'67 GEGENBAUR,** Wir-
belsäule d. Lepid. Jen. Z. iii, 359–414. '93 JUNGERSEN, **H. F. E.,**
Embryonalniere d. Störs, Zool. **Anz. 464,** and (**'94**), Amia, op. cit.
No. 451. '70 KOWALEWSKY, OWSJANNIKOW, U. WAGNER, Stör,
B. Acad. St. Pétersb. xiv, 287–325, **and Mél. Biol. du B. Acad.**
St. Pétersb. vii, 171–183. '91 KUPFFER, K. v., Kopf. v. Acip.
SB. Gesell. Morph. München, **107–123,** and (**'93**) memoir, Leh-
man, München. '90 MARK, **E. L., B. Mus. Comp. Zool. xix,**
1–127. '82 PARKER, **W. K., Skull of Lepid. and Acip. P. Roy.**
S. '87 PELTSAM, **E. D.,** Segmentation **(Russian), MT. Gesell.**
Mosc. Univ. I, Heft **1, and Protocolle d. SB. Zool. Sect.**
Mosc. ('86) I, Heft **1,** 206. **'89 RYDER, J. A., Sturgeon, Am.**
Nat. xxii, 659–660, and ('90) **B. U. S. F. C. viii, 231–281. '78**
SALENSKY, W., Sturgeon, SB. **Gesell. Nat. Kasan ('77), 34**
(Russian). Also Post-Emb. Entwickel **op. cit. ('78) 21** (Russian). (Segmentation) Zool. Anz. **('78)** 243–245, and (Skeleton)
266–269, 288–291. (General) Mém. **S.** Nat. Univ. Kasan, vii
1–226 (Russian). '80 Pt. II, Post-Emb. and Organogeny, op.
cit. **x, 227–545.** Abstract in HOFMAN **and** SCHWALBE'S JB. vii.
213, 217–225. '81 (French), A. de Biol. ii, 233–278.

THE TELEOSTS

(Literature greatly **summarized.**)

GENERAL ANATOMY. — '93 **PARKER, T. J.,** Zoötomy (Cod).
'88 ROLLESTON, Forms **of animal life,** 83–102, and '95 VOGT and
JUNG, Anatomie comparée, Vol. **II. '80** EMERY, C., Fierasfer,
Fauna u. Flora d. Golfes v. Neapel, ii.

SKELETON. — '83 BROOKS, H. S., **Haddock,** P. Roy. S. Dub. iv,
166–196. '90 GILL, T., Skeleton. **notes, P.** U. S. Nat. Mus. xiii,

157-170, 231-242, 377-380. '79 GOETTE, A., Wirbelsäule u. Anhänge, A. mikr. Anat. xvi, 117-142. '84 GÖLDI, E. A. (Derm bones of Catfish, Balistes, Acipenser), Jen. Z. xvii, 401-447. '82 KOSTLER, M., Knochenverdickungen, Z. wiss. Zool. xxxvii, 429-456. '73 VROLIK, A., Verknöckerung, Nied. Arch. Zool. 219-314.

TEETH, INTEGUMENT. — '78 BOAS, J. E. V. (Scarid dentition), Z. wiss. Zool. xxxii, 189-215. '78 CARLET, M., Écailles, Ann. Sci. Naturelle, viii, Art. 8. '86 SCHÄFF, E., Lophobranchier, Inaug. Diss. Kiel.

VISCERA, GLANDS, CIRCULATORY. — '80 BOAS, J. E. V., Conus, Morph. JB. vi, 527-533. '87 BROCK, J., Urogenital, Z. wiss. Zool. xlv, 532-541. '91 CALDERWOOD, W. L., Head kidney, J. Mar. Biol. Ass. ii, 43-46. '82 EMERY, C. (Kidney), A. Ital. Biol. ii, 135-144, Atti. Acc. Rom. xiii, 43-49, ('85) Zool. Anz. viii, 742-744. '77 FÜRBRINGER (Excretory), Morph. JB. iv, 43-49. '83 MAURER, F., Pseudobranchien, op. cit. ix, 229-251. '86 Thymus, op. cit. xi, 129-172. '86 WEBER, M., Abdominalporen (Geschlechtsorgane), op. cit. xii, 366-406.

SWIM-BLADDER. — '90 BRIDGE, T. W., P. Birm. Phil. S. vii, 144-187. '89 BRIDGE and HADDON, A. C., Siluroids, P. Roy. S. xlvi, 309-328, Phil. Trans. ('93) clxxxiv, 65-333. '88 CORNING, H. K., Wundernetz, Morph. JB. xiv, 1-53.

NERVOUS SYSTEM, END ORGANS. — '82 CATTIE, J. T., Epiphysis, A. Biol. iii, 101-196. '88 CHEVREL, R., Sympathetic, C. R. cvii, 530-531. '91 GUITEL, F., Ligne latérale, A. Zool. expér. ix, 125-190, 671-697. '92 HERRICK, C. L., Fore-brain, Am. Nat. xxvi, 112-120, and Anat. Anz. vii, 422-431. '87 LENDENFELD, R. v., Phosphorescent organs, Challenger, xxii, 277-329. '81 MAYSER, P., Gehirn, Z. wiss. Zool. xxxvi, 259-364. '84 SÉDE DE LIEOUX, P. DE, Ligne latérale, Paris, 115 pp.

EMBRYOLOGY, GENERAL. — '91 WILSON, H. V., Sea-bass, U. S. F. C. B. ix, 209-277 (with references). '81-'91 RYDER, J. A., U. S. F. C. R. and B. Larval Teleosts: '77 AGASSIZ, A., P. Am. Acad. v, 117-126, ('78) xiv, pp. 25, ('82) 271-303, and Mem. Mus. Comp. Zool. xiv, 1-56. '87 CUNNINGHAM, J. T., Tr. Roy. S. Edinb. xxxiii, 97-136, ('91) J. Mar. Biol. Ass. ii. '83 HILGENDORF, SB. Nat. Fr. 43-45. '90 HOLT, E. W. L., Sci. Tr. R. Dub. S. 432-474. '80 LÜTKEN, C., Dan. Selsk. xii, 413-613. '91 MCINTOSH, W. C., R. Fish. Scot. ix, 317-342. '90 MCINTOSH and PRINCE, Edinburgh, 4to. '87 RAFFAELE, F., MT. z. Stat. Neap. viii, 1-84, ('90) ix, 305-329.

HERMAPHRODITISM. — '91 Howes, G. B., J. Linn. S. xxiii, 539–558. '67 Jäckel, H., AH. Nat. Gesell. Nürn. iii, 245. '76 Malm, A. W., Œ. v. Ak. Förh. Stockholm. '67 Smith, J. A., P. Roy. S. Edinb. '64 '65, 300–302, ('70) J. Anat. Phys. iv, 256–258. '91 Smith, W. R., R. Fish. Scot. ix, 352. '84 Weber, M., Ned. Tijdschr. Amst. 21–43, ('87) 128–134.

VIVIPAROUS DEVELOPMENT. — '85 Ryder, P. U. S. Nat. Mus. viii, 128–156 (with references).

V. SKELETON OF FISHES

(Cf. Figs. 69, 84, 105, 122, 146, 147, and 310–315)

Cyclostomes.

Vertebral axis a stout notochord; its sheath fibrous, thick, **at the** most with annular hardenings (calcifications). **In** a neural sheath of fibrous cartilage, small (neural) cartilages present. A hæmal fibrous sheath present in tail region. Unpaired fins supported by dermal (?) rays, which take their origin near the vertebral axis, and are braced at the body surface by a longitudinal ligament. Fin rays, usually unjointed, may fork distally.

Cranium a trench **of** fibro-cartilage, at the most but partially roofed; shows **the** cartilaginous elements of the skull of gnathostomes, trabeculæ and parachordals. By some "rudiments" **are** found of "palatoquadrate" and "hyomandibular"; also of **two** branchial cartilages in Myxine, and of "Meckelian cartilage" in Petromyzon. The separate cartilages surrounding the cyclostome mouth forming its frame and supports for tentacles are, moreover, regarded (Pollard) as the homologues of the cirrhial cartilages of Amphioxus; they may well, however, be structures evolved within the Cyclostome group. In Petromyzon a large "basi-hyobranchial" supports the tongue, and a complex "branchial basket" of cartilage is present, which has been compared to the extra-branchial cartilages of shark. **A** "rudimentary branchial basket," as above noted, has been described in Myxine.

	Vertebral Column.	*Fins.*	*Skull.*
Sharks (skeleton cartilaginous).	Notochordal in several forms. Well-developed biconcave vertebræ usually present. Stout, cartilaginous hæmal and neural arches, interneurals. Centra represent the fusion of the surrounding neural and hæmal arches with the sheath; cartilage cells migrate from arches into the sheath at definite points; sheath becomes greatly thickened; ossifications of centra take place concentrically and radially. Ribs rudimentary, the homologues of transverse processes.	Fins supported by basal and radial cartilages; these fuse in the caudal, and often coalesce with the vertebral cartilages. Paired fins similar to the unpaired fins; "monoserial archipterygia." Outer portion of fins supported by parallel unjointed dermal rays. Tail heterocercal. "Claspers" attached to ventral fins of the male.	No centres **of** calcification in the skull; usually no membrane bones. Cranium a cartilaginous trough, modelled on each side into capsules for nose, eye, and ear; a cartilaginous rostrum; a pair of occipital condyles; labial cartilages evolved probably within the group.

Chimæroids (skeleton cartilaginous).	Notochordal; its sheath in anterior trunk region containing many narrow ring-like calcifications; neural arches, with interneurals in anterior trunk region, elsewhere membranous or rudimentary. Cartilage cells of arches migrate broadly into sheath. Ribs rudimentary or wanting.	Unpaired fins generally degenerate. Caudal diphycercal, entirely dermal. Dorsal mainly dermal with radio-basal supports. Dermal fin spine and specialized joint in anterior element. Paired fins shark-like; radials of pectoral fin clustering roundly about distal fin stem (basals). "Claspers" attached to ventral fins of the male.	Highly specialized; brain case, fusing its sides above the brain, rises into a sharp crest; this is continued forward as a vertical plate, separating the eyes and enlarging ventrally in the snout region; a complex series of labial cartilages present; a broad, thin, horizontal floor to the orbital region. Teeth specialized into tritoral areas. Frontal clasping organ in male. Condylar articulation to column.
Lung-fishes Ceratodus (mainly cartilaginous).	Notochordal; sheath with vertebral cartilages; stout neural and hæmal arches, whose cartilage cells, entering the sheath, have fused broadly with it. In the tail region the chord is almost entirely replaced by irregular vertebræ. Ribs present, equivalent to transverse processes of vertebræ.	Unpaired fins continuous into diphycercal tail; their supports radial and basal cartilages. Paired fins "archipterygial." The outer halves of all fins consisting of jointed and bifurcating dermal rays.	Chondrocranium mainly unossified; not separated by condyles from vertebral column. Exoccipitals present. Many investing bones; unpaired (?) frontal and occipital; median ethmoid, large suborbitals, and a pair of nasals. In mouth region large parasphenoid, vomers, palatopterygoid (dentigerous); no maxillary or premaxillary.
Ganoids (cartilaginous and bony).	Usually notochordal; sheath enclosed by hæmal and neural spines. Centra biconcave (in Lepidosteus, however, opisthocœlous). Neural and hæmal elements stout and prominent. Ribs (Lepidosteus) represent transverse processes. Interneurals often present.	Unpaired fins shark-like; dermal portions, however, enlarged; their elements segmented and forked. Paired fins in some forms shark-like (Polyodon); in others, largely dermal, radio-basals constricted, fused, insignificant.	Chondrocranium generally persists and is confluent with vertebral column; massive in Acipenser, with fontanelles in other forms; in all with centres of ossification present. Head surface and mouth surface encased in dermal (membrane) bones.
Teleosts (usually bony).	Notochord early segmented by the ingrowth of the biconcave vertebræ. All ensheathing tissues, arches, processes, ossify; often with intricate fusions *inter se.* The tip of the chord, abruptly upturned, is replaced by "urostyle." Interneurals rarely present. Ribs not homologous with transverse processes; often with accessory parts.	External portion of unpaired fins supported by dermal rays of many types, supporting parts representing mainly radio-basals of older fin types ossified, often with fusions. External portion of paired fins dermal; radio-basals greatly reduced, confluent, ossified. Tail homocercal.	Chondrocranium usually replaced by cartilage bones; often confluent with column. Dermal bones present, to be reduced in general to the type of Amia; often additional scale plates are present in great variety (opercula, branchiostegals).

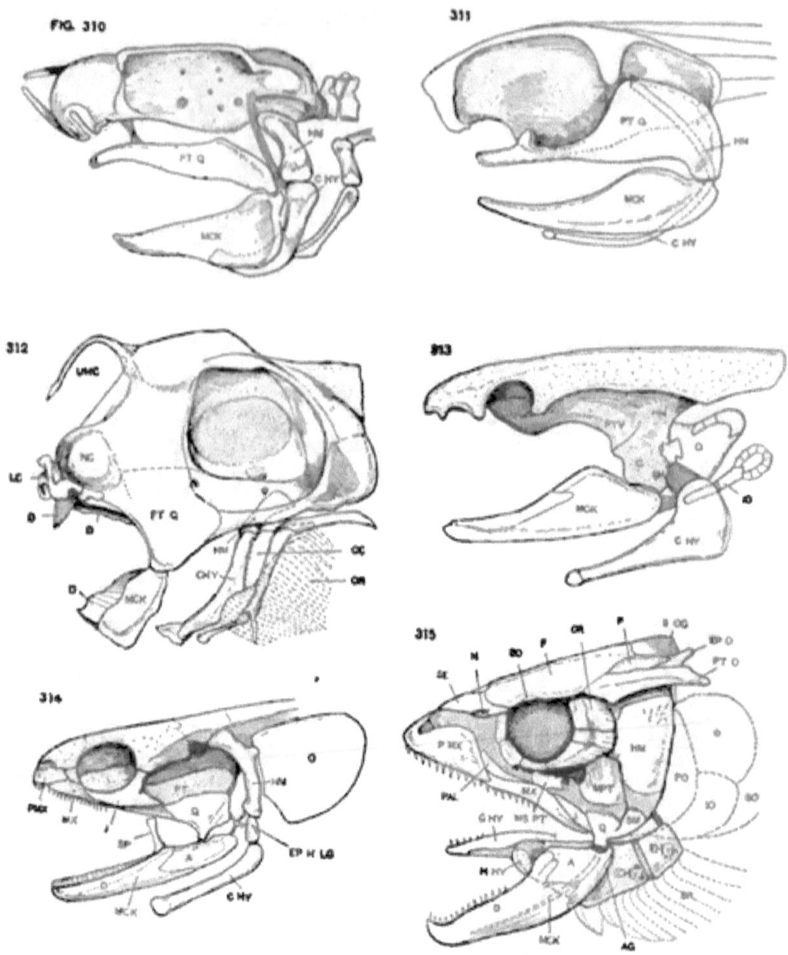

Figs. 310–315.—Skulls of fishes, to illustrate the mode of articulation of jaws and branchial arches. **310.** Skull of *Scyllium*. (After **MARSHALL** and **HURST**.) **311.** *Heptanchus* (Notidanus). (After **HUXLEY**.) **312.** *Chimæra*, ♀ **313.** *Ceratodus*. (Slightly modified after **HUXLEY**.) **314.** *Polypterus*. **315.** Salmon. (After **PARKER**.)

A. Articular. *AG.* Angular. *BR.* Branchiostegal rays. *CHY.* Ceratohyal. *D.* Dentary. *EHY.* Epihyal. *EPH, LG.* Epihyal ligament. *EPO.* Epiotic. *F.* Frontal. *GHY.* Glossohyal. *HHY* Hypohyal. *HM.* Hyomandibular. *IO.* Interoperculum. *J.* Jugal. *LC.* Labial cartilages. *MCK.* Meckel's cartilage. *MPT.* Metapterygoid. *MSPT.* Mesopterygoid. *MX.* Maxillary. *N.* Nasal. *NC.* Nasal capsule. *O.* Operculum. *OC.* Opercular cartilage. *OR.* Suborbital ring. *P.* Parietal. *PAL.* Palatine. *PMX.* Premaxillary. *PO.* Preoperculum. *PTO.* Pterotic. *PTQ.* Palatoquadrate. *PTV.* Palatopterygoid. *Q.* Quadrate. *SOC.* Supraoccipital. *SE.* Supra-ethmoid. *SM.* Symplectic. *SO.* Supraorbital. *SP.* Splenial. *UMC.* Upper median cartilage (not frontal spine of male).

Figs. 310, 314, 315 are regarded by HUXLEY as "hyostylic" (*i.e.* the hyoid element, *HM*, attached by ligaments to the jaw hinge, taking an important part in the suspension of the jaw; **311**, a modified hyostylic condition; the hinder upper margin of *PTQ* becoming greatly enlarged, and attached by ligaments to the skull, is spoken of as "amphistylic"; **312–313**, were "autostylic," *i.e.* the upper jaw element fused with the skull.

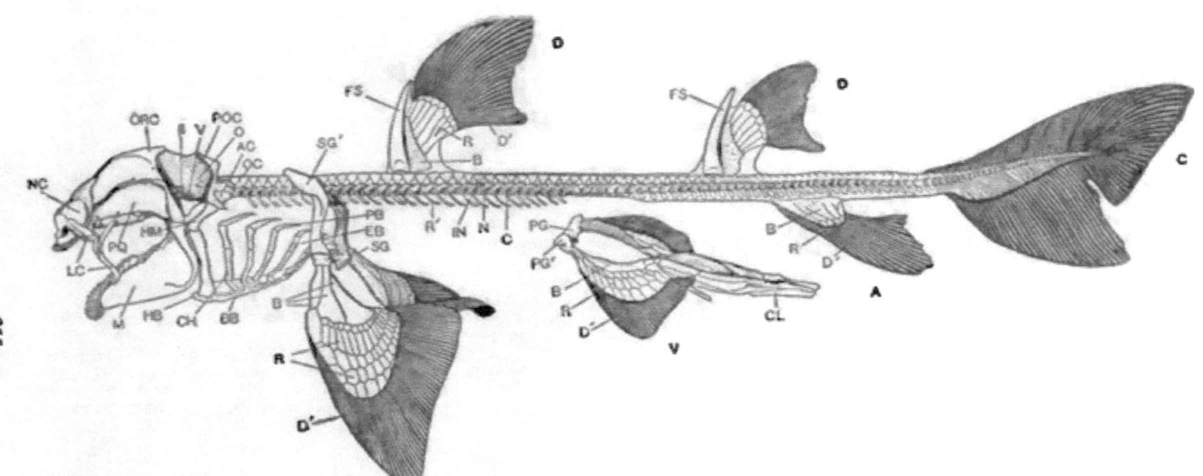

Fig. 84. — Skeleton of shark, *Cestracion galeatus* (♂). (From preparation of H. A. WARD. Drawing by ARNOLD GRAF).

A. Anal fin. *AC.* Auditory capsule. *B.* Basal cartilaginous fin supports. *BB.* Basibranchial. *C.* Caudal fin. *CH.* Ceratobranchial (the gill arch element immediately anterior is the ceratohyal). *CL.* Claspers. *D.* Dorsal fins. *D'.* Dermal rays of fin. *EB.* Epibranchial. *FS.* Fin spine. *HB.* Hypobranchial. *HM.* Hyomandibular. *IN.* Interneural plate. *N.* Neural process. *NC.* Nasal capsule. *O.* Orbit. *OC.* Occipital condyle. *ORC.* Orbital crest. *P.* Pectoral fin. *PB.* Pharyngobranchial. *PG.* Pelvic girdle. *PG'.* Dorsal process of *PG.* *POC.* Postorbital ridge. *PQ.* Palatoquadrate. *R.* Radial. *R'.* Rib. *SG.* Shoulder girdle. *SG'.* Dorsal process of *SG.* *V.* Ventral fin. *II.* Foramen of optic nerve. *V.* Foramen of fifth cranial nerve.

VI. RELATIONS OF THE JAWS AND BRANCHIAL ARCHES OF FISHES

(Cf. Figs. 310–315)

The nasolabial cartilages of sharks and Chimæroids may represent a preoral branchial arch. What is generally spoken of as the first (or mandibular) arch consists on each side of two elements — the upper or Palato(-pterygo)-quadrate, the lower or Meckelian cartilage. Each half of the second (or hyoid) arch consists of an upper element, Hyomandibular = suspensorium, and a lower, Ceratohyal. Each of the hinder gill arches, from 4 to 6 in number, consists on either side of three segments; to these are added one or more ventral segments for inter-attachment in the median ventral line.

	Mandibular.	*Hyoid.*	*Branchial Arches.*
Shark, Type Scyllium **or** Raja (*Hyostylic* *).	Palatoquadrate functions as upper, and Meckelian as lower jaw; their oral margins alone of all mouth parts dentigerous; teeth of shagreen pattern. The palatoquadrate is loosely attached to the cranium by ligaments, anterior or posterior, as seen in the figure.	Hyomandibular serves to suspend the hinge of the jaw; its attachment proximally to otic skull region and distally to jaw hinge, stout, ligamentous. Ceratohyal at its distal end supports anteriorly the ventral ends of the hinder branchial arches; on ectal rim may bear rod-like opercular cartilages (branchiostegals).	Proximal segment flat, spatulate, spreading out in the fibrous tissue between the gullet and vertebral column. Median segments may bear thin, plate or rod-like supports for the dermal gill flaps.
Shark, Type Heptanchus (*Amphistylic* *).	As above; the hinder portion of the palato-quadrate, however, greatly enlarged and more perfectly attached by ligaments to the cranium. It broadly overlaps the hyoid arch. (Cestracion gives a type intermediate between this and Scyllium.)	The second arch much reduced, rod-like, owing to the wide development of the hinder portion of the first arch. Hyomandibular can have little function as suspensorium, and the more perfect ligamentous connections ventrally have greatly reduced the size of the ceratohyal.	**As above.**
Chimæroids (*Autostylic* *).	Palatoquadrate functions as upper jaw, although by fusion with the anterior cranial cartilage its limits are obscure; its caudal margin is best marked. Meckelian stout and short. The homologies of the dental plates are obscure, although suggesting those of Cochliodont sharks.	Hyomandibular no longer the suspensorium; entirely homologous with the proximal element of the third arch. Ceratohyal as in sharks; bears at ecto-proximal margin an " opercular " cartilage, bifid or trifid; from this cartilage the branchiostegal rays (to whose fused bases it owes its origin) pass distad to the lower portion of the ceratohyal.	Similar to sharks.

Dipnoan, Ceratodus (*Autostylic* *).	Palatoquadrate cartilage functions as upper jaw; its mesial border fused with chondrocranium, and bearing anteriorly the dental plate; quadrate element imperfectly ossified. Meckelian cartilage flattened, ensheathed on ectal surface with dentary and angular; membrane bones (removed in the figure); ensheathed on ental surface with splenial.	Hyomandibular cartilaginous, fused proximally with chondrocranium, anteriorly with palatoquadrate; at its distal end (?) a new element now appears, the symplectic, forming a hinder process of the quadrate. An opercular apparatus (membrane bones =metamorphosed scales) is attached by ligaments to the caudal margin of the hyomandibular. Ceratohyal, encased in a membrane bone, is attached by ligaments to suspensorium and mandible.	Cartilaginous; segments not prominent. Ventral elements rudimentary. Fifth arch appears unjointed, appended to fourth.
Teleostome, Polypterus (*Hyostylic* *).	Palatoquadrate fused at mesial border with chondrocranium. In it the following membrane bones appear: quadrate, pterygoid, mesopterygoid, metapterygoid, and anteriorly a small palatine element. The dentigerous portion of the upper jaw is now composed of membrane bones arising near the mouth rim, in front of the palatoquadrate, premaxilla, maxilla, vomer, and jugal. Meckelian cartilage, reduced in size, is encased in membrane bones, dentary, articular, and splenial.	Hyomandibular distinct and partly ossified; connected proximally by ligaments with otic region of skull, distally with quadrate, mandible, and epihyal (=symplectic (?)). Opercular apparatus appended to caudal margin. Ceratohyal a cylindrical membrane bone.	Partly ossified; segments prominent; basal ends of arches produced forward and broadly fused.
Teleostome, Salmon (*Hyostylic* *).	Palatoquadrate not widely different from Polypterus; membrane bones more sharply differentiated. Meckelian cartilage reduced to a rod-like rudiment, encased in stout membrane bones; an additional ossification is present — the angular — and in kindred fishes several additional ones may occur.	Hyomandibular stout, firmly articulated with skull; cephalad, it is connected with pterygoid elements, the ossification at its distal end — symplectic — firmly cementing it to the quadrate; its proximal end is densely ossified, and to this region are attached the opercular plates. Ceratohyal (of three segments) is attached by a stout ligament to the hyomandibular.	Ossified; compressed antero-posteriorly.

* These terms, as used by Huxley (*Pro. Zool. Soc.*, '76, pp. 40-44), may be defined as follows: —

Autostylic, where " the *Mandibular arch* is attached directly to the skull by that part of its own substance which constitutes the suspensorium."

Hyostylic, where the hinge of the jaw is mainly supported " by the dorsal element of the hyoidean arch, or hyomandibular."

Amphistylic, where the " palatoquadrate cartilage is quite distinct from the rest of the skull, but is wholly, or almost wholly, suspended by its own ligaments, the hyomandibular being small and contributing little to its support." The entire problem of the suspensorium of the jaws of fishes seems at present in an exceedingly unsatisfactory condition, especially in view of the results recently maintained by Pollard ('94, *Anat. Anz.*, Vol. X., pp. 7-25). In Figs. 310-315, the present writer will be seen to incline to the older view, but cannot regard the autostylism of Chimæroid as equivalent to the autostylism of Dipnoan.

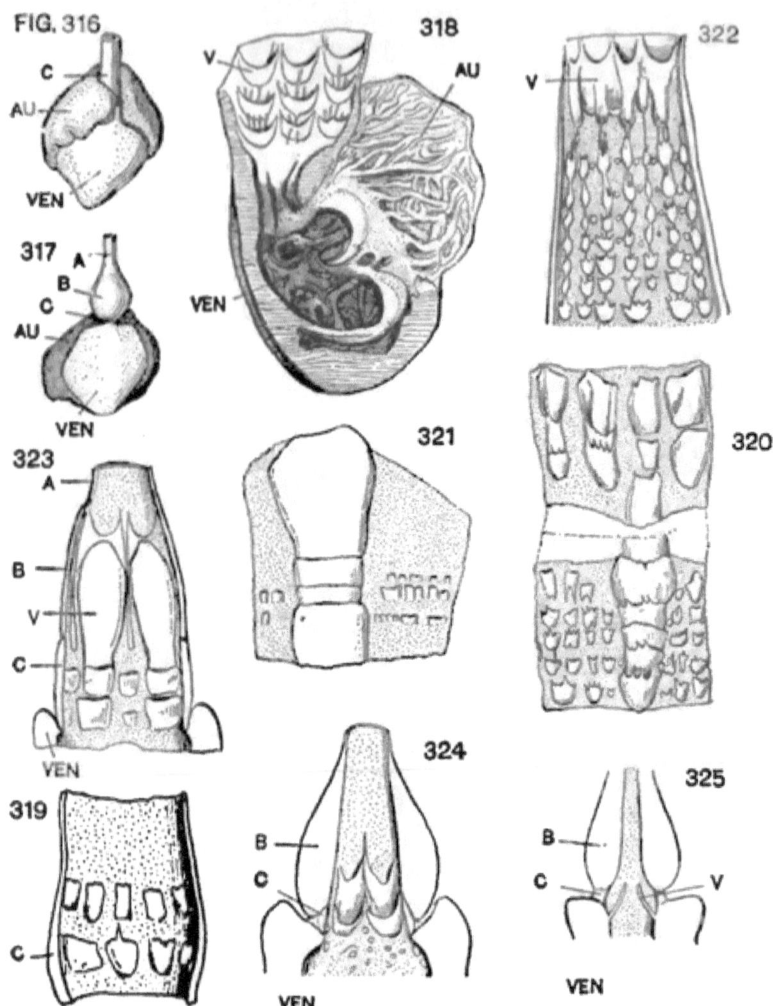

Figs. 316–325. — Heart and arterial cone of fishes. 316. Heart of shark; 317, Heart of catfish, *Silurus glanis*; 318, Heart of shark, shown opened at the side. 319. Conus arteriosus (inner view) of *Chimæra*. 320. Conus of *Ceratodus*. 321. Conus of *Protopterus*. 322. Conus of *Lepidosteus*. 323. Conus of *Amia*. 324. Conus and bulbus of the Teleost, *Butrinus*. 325. Conus and bulbus of the Teleost, *Clupea*. (Figs. 316–318 after WIEDERSHEIM, 320–325 after BOAS.)

A. Aorta. *AU.* Auricle. *B.* Bulbus. *C.* Conus. *V.* Valves. *VEN.* Ventricle.

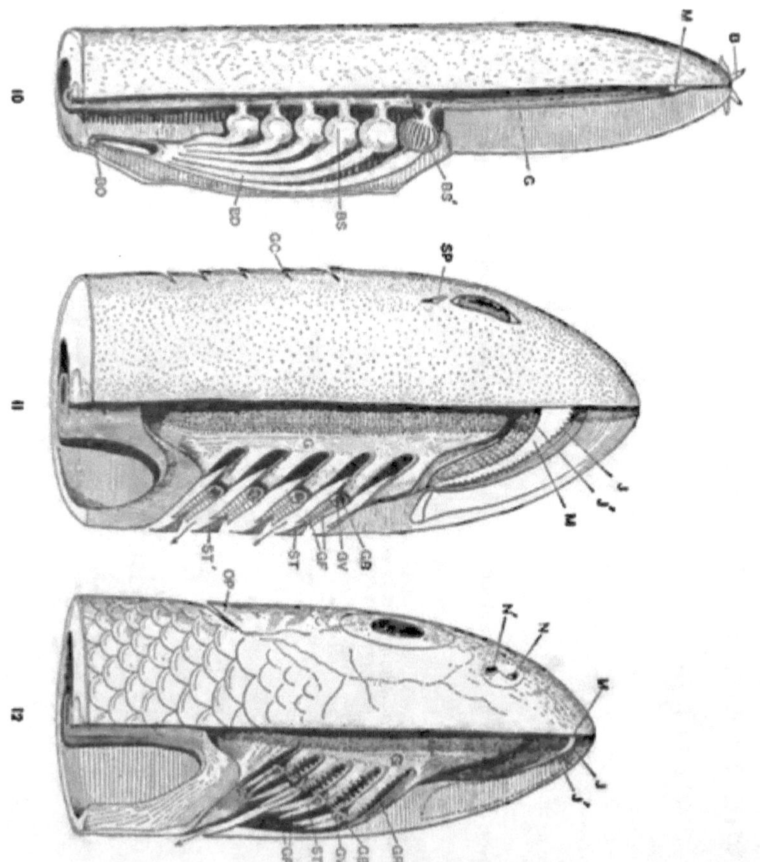

Figs. 9–12.—Arrangement of gills of Bdellostoma (9), Myxine (10), Shark (11), and Teleost (12). In each figure the surface of the head region is shown at the left.

B. Barbels. *BD.* Outer duct from gill chamber. *BS. BO.* Common opening of outer ducts from gill chambers. *BS.* Branchial sac, or gill chamber. *BS'.* Branchial sac, sectioned so as to show the folds of its lining membrane. *G.* Lining membrane of gullet. *GB.* Gill bar, supporting vessels and filaments of gills. *GC.* Outer opening of gill cleft. *GF.* Gill filament. *GR.* Gill rakers. *GV.* Vessels of gill. *J, J'.* Upper and lower jaw. *M.* Mouth opening. *N, N'.* Anterior and posterior opening of nasal chamber. *OP.* Operculum. *SP.* Spiracle. *ST.* Tendinous septum between anterior and posterior gill filaments. * Denotes the inner branchial opening; →, the direction of the water current.

259

VII. THE HEART OF FISHES

(Cf. Figs. 316–325)

It consists of sinus venosus, atrium, ventricle, with either conus arteriosus or bulbus.

Cyclostomes. Conus non-contractile, without valves.   The pericardium a continuation of the peritoneum.

Sharks. Conus contractile, with from 3 (Zygæna) to 7 (8) transverse rows of valves (Chlamydoselache). Pericardium connected with cœlome by bifurcating canal.

Holocephali. Conus short, almost converted into a bulbus, with 3 rows of valves. Pericardium communicating with cœlome by single aperture.

Dipnoans.
Protopterus. Conus twisted, with 4 rows of valves. Aperture of pericardium to cœlome rudimentary. Sinus venosus and atrium divided longitudinally.

Ceratodus. As in Protopterus; 8 rows of valves present.

Teleostomes.
Lepidosteus. Conus long and straight, with 8 (to 11) rows of valves. Pericardium separated from cœlome.

Sturgeon. As in Lepidosteus, but with 3 rows of valves. Pericardium communicating with cœlome.

Amia. Conus is short, an anterior thickening (bulbus) appearing; 3 rows of valves. Pericardium separated from cœlome.

Butrinus Conus is much contracted proximad, anterior thickening (bulbus) large; 2 rows of valves. Pericardium separated from
(Clupeoid). cœlome.

Teleosts. Conus is practically wanting; bulbus large; a single pair of valves between ventricle and arterial trunk. Pericardium separated from cœlome.

VIII. A COMPARISON OF GILLS, SPIRACLE, GILL–RAKERS, AND OPERCULA

(Cf. Figs. 9–12 and 326–331)

Cyclostomes. The gill is a dilation of the tubular duct passing outward from the gullet; its vascular surface is folded longitudinally (i.e. ecto-entad). In Bdellostoma there may be 13 or 14 pouches on each side. In Myxine 6 pairs are present (with a rudimentary seventh cleft, the œsophageal duct, on the left side), having a common outer opening. In Petromyzon 7 pairs open directly to the body surface — from the gullet in the larva, and from a diverticulum of the gullet, the branchial sac, in the adult.

	Spiracle.	Hyoidean Gill.	Gills.	Gill Defences.
Fishes.	Each gill consists of an upright row of horizontal plate-like vascular filaments; it may be attached to the anterior or posterior side (or both) of the ectal rim of a gill bar; its filaments thus protrude into the gill slit, and are constantly bathed in the outgoing water. Gill rakers are antero-posterior projections of the ental rims of the gill bars, which interlock when the gills are drawn together, screening the gill openings and preventing the escape of food.			
Sharks.	Present with pseudobranch; spiracular opening sometimes greatly reduced **or** absent.	Only on posterior side of hyoid.	5–7, with filaments on both sides of gill bars; gills decrease little in size hindward; in some forms (Selache) a screening apparatus of gill rakers is developed.	Dermal folds produced at the anterior margin of each cleft; this is sometimes supported by cartilaginous rods.
Chimæra.	——	"	4, but the last bar with filaments only on **its** anterior side; an imperfect screening apparatus present.	A stout dermal fold encircling all gill arches, supported by cartilaginous plates and rods.
Ceratodus.	——	——	4 complete gills; **closely interlocking gill** rakers.	Operculum, pre- and interopercula.
Protopterus.	——	Only on posterior side of bar (opercular gill).	**4 complete gills; filaments lacking, however, on upper halves of the gill arches.**	Operculum **and interoper**culum.
Polypterus.	Present; no pseudobranch.	——	4, but **the last with filaments** only on anterior side.	Operculum, pre- **and sub**opercula.
Lepidosteus.	Rudiment; pseudobranch present.	Only on posterior side of bar (opercular gill).	**4 complete gills.**	Operculum, **pre-, sub-, and** interopercula.
Acipenser.	Present; also pseudobranch.	"	"	"
Polyodon.	"	No opercular gill.	"	"
Amia.	(Rudiment of spiracle — Wilder.) No pseudobranch.	Opercular gill.	"	"
Teleosts.	——	Rudiment on posterior face.	" (Or less, *e.g.* Malthe, 2½ gills.)	⸱

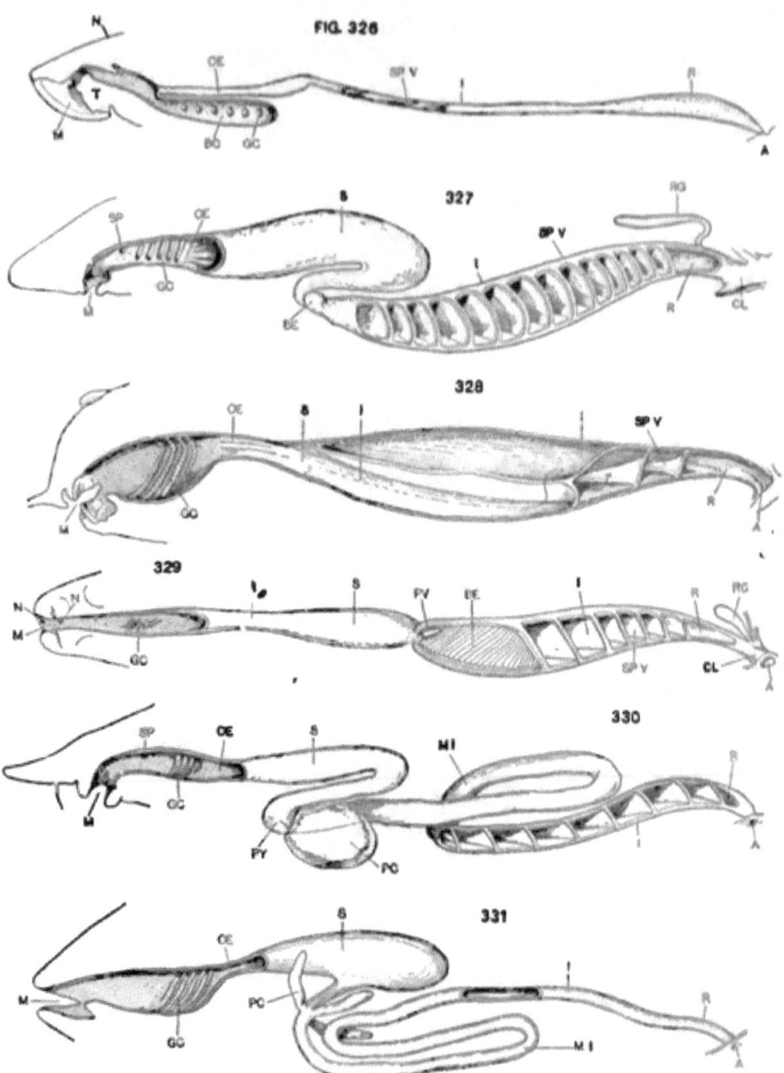

Figs. 326–331.—Digestive tracts of fishes. 326. Cyclostome, *Petromyzon*. 327. Shark. 328. Chimæroid, *Callorhynchus*. 329. Lung-fish, *Protopterus*. (After W. N. PARKER.) 330. Ganoid, *Acipenser sturio*. 331. Perch. (After WIEDERSHEIM.)

A. Anus. *BC*. Branchial chamber. *BE*. *Bursa entiana* (duodenum). *CL*. Cloaca. *GC*. Gill openings. *I*. Intestine. *M*. Mouth. *MI*. Mid-gut. *NN''*. Anterior and posterior nares. *OE*. Gullet. *PC*. Pyloric cœca (pancreas). *PY*. Pyloric end of stomach. *R*. Rectum. *RG*. Rectal gland. *S*. Stomach. *SP*. Spiracle. *SP. V*. Spiral intestinal valve.

IX. DIGESTIVE TRACT

(Cf. Figs. 326–331)

The mouth region of all forms lacks salivary **glands.**

Cyclostomes. Straight, little dilated; rudimentary spiral **valve in** intestine. A short mesentery in hindmost part.

Sharks. Wide gullet, siphonal stomach, short intestine provided with spiral valve (having as many as **40 turns**); short rectum with digitiform cæcum; cloaca. **Mesentery at** beginning and end of digestive tube. Spleen **often broken** into lobules. **Gullet** pierced with 5 (6 or **7**) **gill clefts**; spiracle present, its gill supplied by first branchial **artery. An** opercular skin **fold present in Chlamydoselache.**

Holocephali. **Short, wide,** almost straight, **no marked stomachic** dilation; intestine with spiral valve of 3 turns, no cloaca. Dorsal mesentery reduced to 4 string-like supports. **No** spiracle; **gill arches 5,** compressed; outwardly protected by opercular skin fold, with rudimentary **cartilaginous supports.**

Dipnoans.

Ceratodus. **Short;** narrow gullet, tubular stomach, **intestine with spiral valve of 9** turns, short cloaca with **diverticulum.** Dorsal mesentery greatly reduced. **No spiracle; 5 gill arches; hyoidean gill.**

Protopterus. **Short, almost** straight; stomach small; **intestine with spiral** valve of 9 **turns,** short cloaca **with dorsal diverticulum. No** spiracle, 5 gill arches, of which the hinder 3 present external branchiæ.

Teleostomes.

Polypterus. Short, straight; a single pyloric cæcum, **spiral valve of 8 turns, no cloaca, single (or double) rectal cæcum. A spiracle. Teeth** labyrinthodont.

Lepidosteus. Large, long stomach, recurved at end; **a mass of pyloric cæca;** short, twisted, small intestine; short, large intestine, with spiral valve of 3 (rudimentary) **turns. No cloaca. Well-marked** dorsal mesentery. Teeth labyrinthodont.

Acipenser. Large siphonal stomach; **pyloric cæca in fused mass (but separate in** Polyodon); short, small intestine; large intestine, with spiral valve of **8–9 turns; no cloaca. Dorsal mesentery largely fenestrated.** Spiracle present.

Amia. Short, wide, pouch-like stomach, no pyloric cæca; long, stout, small intestine, little curved; short, stout, large intestine, with spiral valve of 4 turns; no cloaca. Dorsal mesentery at hind gut largely fenestrated.

Teleost. Digestive tract of many types; stomach usually recurved; **many pyloric** cæca; intestine long and narrow (with spiral valve **only in Cheirocentrus), with little difference between** anterior and posterior divisions; no cloaca. Many conditions of mesentery. **Spleen single. No spiracle. Accessory** branchial apparatus.

X. SWIM–BLADDER

(Cf. Figs. 13–19)

Cyclostomes.	Wanting.
Sharks.	Wanting. (Cæcal pouches of pharynx perhaps homologous.)
Holocephali.	**Wanting.**
Dipnoans.	
Protopterus.	**Paired; opens at** glottis on **ventral** side of pharynx; its lining membrane thrown into **irregular ridges ; its vascular supply** from true pulmonary artery arising from the point of union **of efferent branchial artery, thence returns by vein to left** sinus venosus.
Lepidosiren.	As in Protopterus.
Ceratodus.	**Azygos; opens at glottis on right ventral side of pharynx; lining membrane in regular ridges;** vessels similar to **Protop-terus ; pulmonary artery, however, arising from the fourth efferent branchial artery.**
Teleostomes.	
Polypterus.	**Paired; opens on** ventral wall of pharynx; glottis **present; vascular supply from paired** 4–5 **branchial arteries, returns by systematic veins** (portal ?).
Lepidosteus.	**Azygos; opens to** dorsal wall of gullet; vascular supply from dorsal **aorta (? cœliac artery) ; lining** membrane regularly **ridged.**
Amia.	**Azygos; opens to dorsal wall of gullet; vascular supply from** paired fourth branchial **artery, returns by systematic veins; lining membrane irregularly ridged.**
Sturgeon.	**Azygos; opens without glottis to dorsal wall of pharynx; lining membrane smooth; vascular supply from dorsal aorta returns by systematic veins.**
Teleosts.	**As in sturgeon.** Often subdivided, sometimes paired, **sometimes with many diverticula; its long air duct may connect with the side of the** gullet (Erythrinus) **; connects dorsally with the throat in Physostomes; its opening obliterated in Acanthopterygians,** Anacanths, Lophobranchs; **may be sound-producing; connected with the ear in Cyprinoids and Siluroids, and in the latter** group it may **be compressed at will. Its substance may be calcified —Cyprinoids, Siluroids,** and in fossil Crossopterygians. May **be absent (Pleuronectids).**

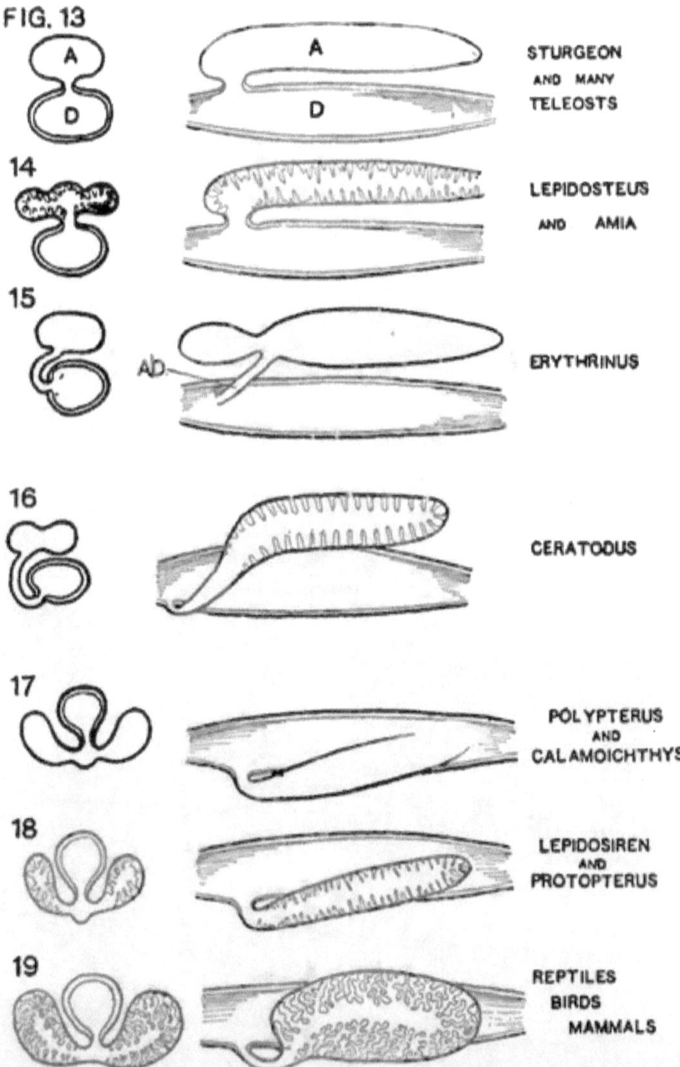

Figs. 13-19. — Air-bladder of fishes, shown from the front and sides. Cf. p. 264. *A.* Air- or swim-bladder. *AD.* Air duct. *D.* Digestive tube. (After WILDER.) 13. Sturgeon and many Teleosts. 14. Amia and Lepidosteus. 15. Erythrinus, a Cyprinoid Teleost. 16. Ceratodus. 17. Polypterus and Calamoichthys. 18. Lepidosiren and Protopterus. 19. Reptiles, birds, and mammals. The diagrams illustrate the paired or impaired character of the organ, its varied mode of attachment to the digestive tube, and the smooth or convoluted condition of its lining membrane.

XI. GENITAL SYSTEM

(Cf. Figs. 332–337)

Cyclostomes.
Myxine and Bdellostoma.
Hermaphrodite; ova and sperm, however, ripened at different periods (of life ?); the younger adults are thus (Nansen, Cunningham) males, the older, females (=Protandric condition). Genital products fall from gland to body cavity, thence out by abdominal pores. Eggs fasten together at ends by their horny processes, **and** appear to be emitted in a string. Fertilization unknown; perhaps occurring in abdominal cavity.

Petromyzon.
Sexes separate. Eggs and sperm fall into body cavity, thence emitted through abdominal pores. **Fertilization takes place** in water.

Shark.
Sexes separate. Testes paired, rounded; sperm passes **outward** through much convoluted vas deferens **and** vesicula seminales (= Wolffian duct) to sperm sac and urinogenital sinus; through urinogenital papilla and grooves of claspers into cloacal opening and oviduct of female. Eggs pass from paired ovaries to cœlome, thence to common opening of paired oviducts (Müllerian), are at once fertilized and receive shell (except in Læmargus; in Trygonorhina 2 eggs to 1 shell) in upper oviduct, thence pass outward through dilated uterus-like portion of oviduct by unpaired aperture, opening behind ureter, to cloaca. In viviparous sharks the walls of vitelline sac fuse with oviduct, establishing " placental " circulation.

Chimæroid.
Similar to oviparous shark.

Dipnoan.
Ceratodus.
Sexes separate. Sperm passes outward through sperm duct (Wolffian ?) and dilated terminal portion of segmental duct through cloaca and vent. Eggs drop from ovaries to cœlome, thence enter paired openings of oviducts (Müllerian) and escape through a common opening into cloaca : are fertilized in the water.

Teleostomes.
Sturgeon.
Amia.
Lepidosteus.
Teleosts.
Similar to Dipnoan, genital opening separate behind anus.
  "   "   "   "   "   "   "
  "   "   "   "   "   "   "

Sexes separate in nearly every instance. Sperm and ova pass outward through cœlome and genital tubes, vasa deferentia and oviducts, to unpaired genital pore. Oviduct and vas deferens may be secondary structures. (Oviduct continuous with ovarian capsule, although there appear many and noteworthy modifications.) A vas deferens (Wolffian ?), which may fuse more or less with its fellow, sometimes opens into or in front of ureter. An intromittent organ (modified anal fin spine or elongated urinogenital papilla) may sometimes be present. A single oviduct is often present, opening in front of ureter. Ovary sometimes enclosed in sac continued into the oviduct—this dilates in the case of viviparous fishes (Embiotocidæ), and is widely specialized for attachment and emission of young. Ovaries or testes (occasionally single) paired, may be confluent. Cod, flat-fish (Pleuronectid), and herring may exceptionally be diœcious, sometimes testis on one side, ovary on the other. Some forms (Serranus) appear to be regularly hermaphrodite.

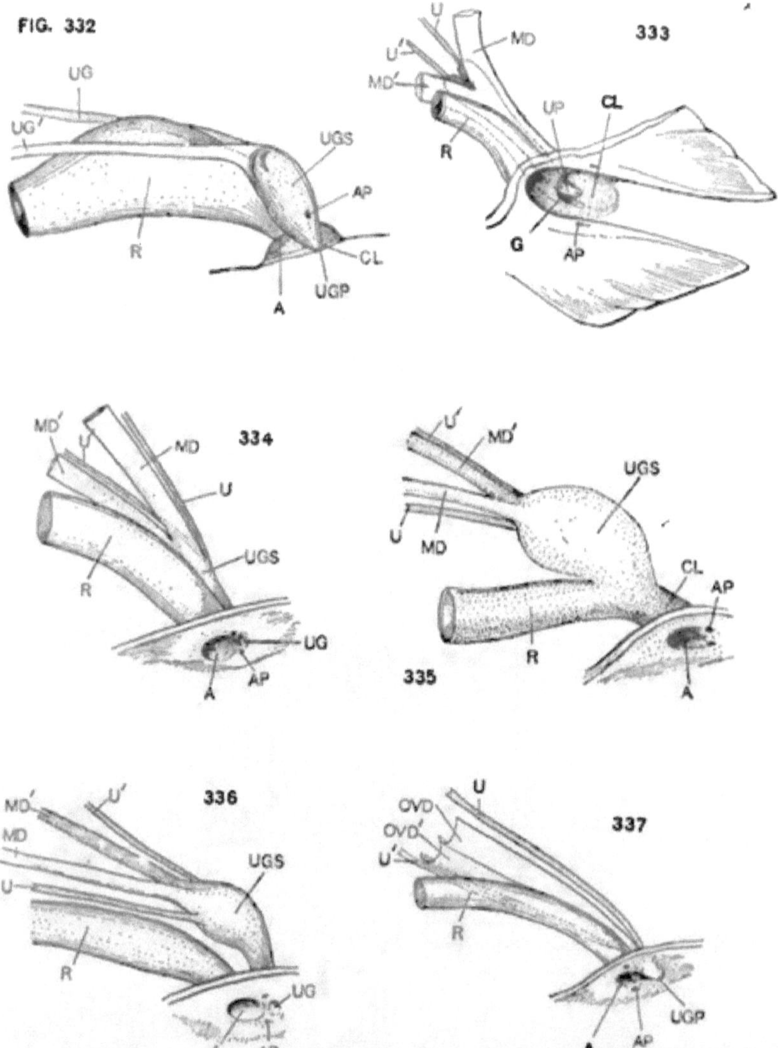

Figs. 332-337.—Urinogenital ducts and their external openings. 332. Cyclostome, *Petromyzon*. (After W. K. PARKER.) 333. Shark, ♀. 334. Chimæroid, juv. ♀. 335. *Ceratodus*. 336. Ganoid, ♀. 337. Teleost (Salmonoid), ♀. (After BOAS.)

A. Anus. *AP*. Abdominal pore. *CL*. Cloaca. *G*. Genital opening. *MD, MD'*. Left and right Müllerian ducts. *OVD, OVD'*. Left and right oviducts (not Müllerian ducts). *R*. Rectum. *U, U'* Left and right ureters. *UG*. Urinogenital opening. *UG', UG''*. Left and right urinogenital ducts. *UGP*. Urinogenital papilla, showing distal opening. *UGS*. Urinogenital sinus. *UP*. Urinary papilla.

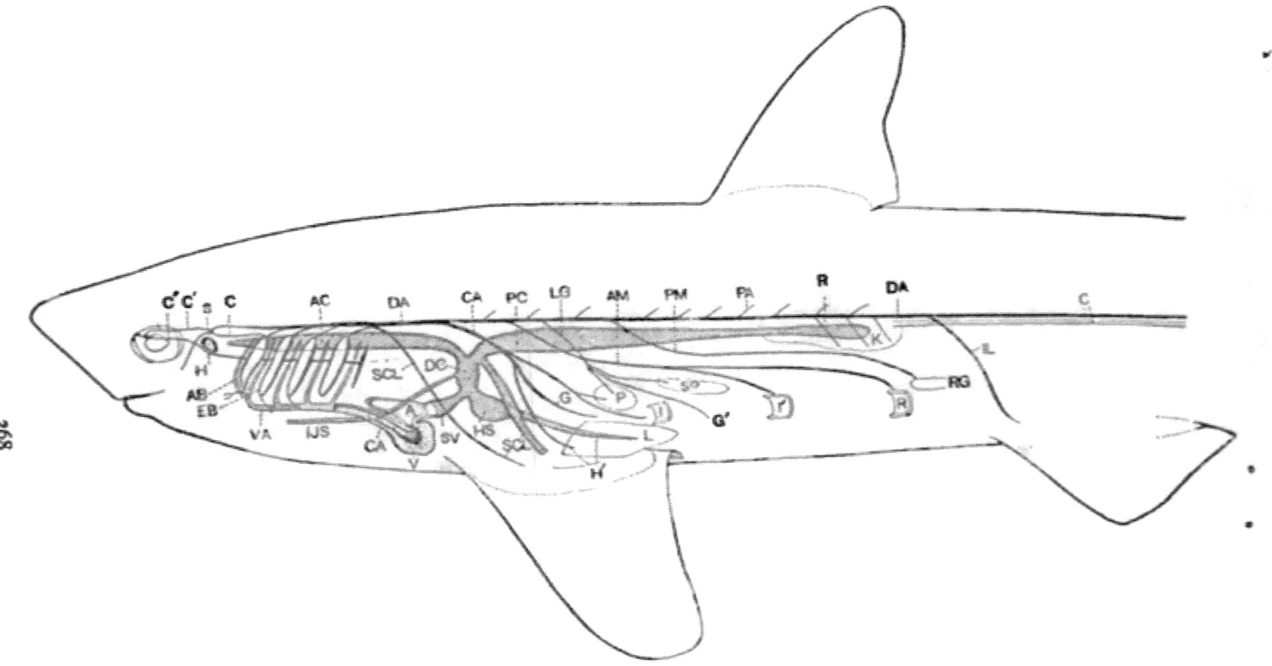

C" C' S C
AC
DA
CA
PC
LG
AM
PM
PA
R
DA
C
H
AB
EB
VA
IJS
CA
V
SV
HS
SCL
A
DC
SCL
G
P
SO
G'
L
H
IL
RG
R
K

XII. CIRCULATION IN FISHES

The course **of** the blood supply of fishes may be traced in outline (taking as a type an Elasmobranch) in the diagram, Fig. 338.

Impure blood from every part of the body reaches the heart (atrium) through the *sinus venosus* (*SV*), which is in turn fed by greatly dilated veins — paired *jugular* (*JS*), paired *Cuvierian* (*DC*), and azygos *hepatic* sinuses (*HS*) ; a Cuvierian sinus again is fed by *anterior* and *posterior* (*cardinal*) *caval sinuses* (*AC* and *PC*), and a *brachial* (*subclavian*) vein. Each **vein** sinus is the collecting vessel of its peculiar region ; thus the jugular vein collects the impure blood from the region of the throat ; hepatic from capillaries of the liver, through which comes the blood from intestine and spleen (hepatic-portal system) ; *anterior cardinal* from the **dorsal** head region ; *posterior cardinal* from the kidneys, through which the blood from the hinder body region (through the caudal region) **has been** passed (venal-portal system).

Venous blood arriving at the heart is now to be aerated at the gills ; it passes from *auricle* (atrium) to *ventricle*, thence to **the *afferent* aortic trunk** (*AA*) ; into its 5 pairs of branches (*EB*), and into the gill lamellæ.

Purified in the gill capillaries, the blood collects again into efferent branchial vessels **to** be passed to **all parts** of the body ; accordingly, from each efferent branchial loop (*AB*) a vessel passes mesad, uniting with its fellows in the **median** line ; **thus** is formed the dorsal aorta (*DA*) which supplies pure blood to all body regions, except in part the head. The head region **receives** its arterial blood largely from the foremost efferent branchial loop ; a carotid artery (*C*) arises dorsal, dividing anteriorly into internal and external (*C'C''*) branches ; the former fusing anteriorly with **its** fellow of the opposite side completes the cephalic blood circle ; a *hyoidean* artery (*H*) arises somewhat ventral, supplies the spiracle (*S*), and fuses anteriorly with the internal carotid. The hyoidean is accordingly to be regarded as a rudiment of a foremost branchial loop.

The arterial supply **of** the trunk region is entirely derived from the dorsal aorta ; **the** *subclavian* (*SCL*) supplies the pectoral region, the *cœliac axis* (*CA*) through its branches the anterior viscera — liver, anterior stomach, intestine, and pancreas, the *anterior mesenteric* (*AM*) intestine and rectum ; the *lieno-gastric* (*LG*) the pancreas, stomach, and spleen ; the *posterior mesenteric* (*PM*) the rectal gland ; and of the paired vessels the *renals* (*R*) supply the kidneys, the ***parietals*** (*P*) the body walls, and the **large** *iliacs* (*IL*) the pelvic region. A *caudal* artery, the terminal portion of the dorsal aorta, **is** the supply **trunk of** all parts behind the visceral **cavity.**

COMPARISON **OF THE CIRCULATORY VESSELS IN OTHER GROUPS**

Chimæroids. Vessels essentially similar to those of **Elasmobranch. On each axillary** (*subclavian*) **artery is a vascular bulb connected** problematically with the sympathetic system.

Dipnoans. Arterial conus shortened by its twisted course ; afferent aorta much shortened, the four branchial vessels appearing to arise
Protopterus. from the same point (as in Amphibia) ; branches of the hinder 3 supplying the external branchiæ. No cephalic circle. Efferent branchial vessels collected on each side into a single trunk before fusing **with** the dorsal aorta ; from each collecting trunk arises a pulmonary artery. Veins scarcely larger in diameter than arteries, their relations to the sinus venosus differing little from Elasmobranch. The right posterior cardinal, however, is anomalous ; it receives in **the** region of the liver the hepatic veins, and appears, therefore, to represent as well the hepatic sinus. (According to **W. N.** Parker it is the post-caval of higher vertebrates, its fellow and the right posterior cardinal not occurring.) Pulmonary **veins** unite before entering the small left atrium, the aerated blood passing thus into the grooved ventricle, and thence into the **grooved** conus, to be delivered into the first **3** arches almost unmixed with venous blood. Lymphatic vessels apparently wanting.

Teleost. Conus greatly reduced, **bulb** prominent, last branchial arch arising close to **its** tip. **First** afferent vessels **absent ; efferent**
Cod. branchial **vessels** represented **by** single tubes instead of loops, the two hindmost pairs unconnected ventrally, but **all** are **connected** dorsally, forming at their tops **a** horizontal efferent vessel. This fuses with its opposite fellow both cephalad and caudad, **the latter** fusion giving rise **to the** dorsal aorta. The principal visceral arteries, however, do not arise from the dorsal aorta, but **anteriorly** from the trunk **of** the efferent branchial vessels ; the gastric artery is the only large ventral branch of the dorsal aorta. **Veins** resemble arteries in size ; Cuvierian sinus-ductus has now reduced to an elongated and tubular vessel, and the hepatic **veins** open directly to the sinus venosus.

XIII. EXCRETORY SYSTEM AND URINOGENITAL DUCTS

(Cf. Figs. 332–337)

	Pronephros.	Mesonephros.	Metanephros.	Segmental Duct.	Müllerian Duct.	Wolffian Duct.	Remarks.
Bdellostoma.	Persists (?).	Is greater part of functional kidney.	—	Remains undivided.	—	—	Segmental duct serves as ureter. In *M.* ureter terminates in sunken urogenital sinus. In *P.* ureter is wide, coalescing near termination with urogenital papilla.
Myxine.	" "	" "	—	" "	—	—	
Petromyzon.	Trace of pronephros of larva persists.	Functions ; retains nephrostomata.	—	Splits into Müllerian and Wolffian ducts (?).			
Sharks. Male.	Well marked in embryo.	Functions (Epididymis).	Kidney.*	Splits into Müllerian and Wolffian ducts.	Rudiment.	Vas deferens, urinogenital sinus and sperm sac.	**Ureters are confluent ducts of " metanephros " (along its border usually separate).**
Female.	" "	Rudimentary (Parovarium).	" "	" "	Oviduct.	Small anteriorly (duct of mesonephros), dilates behind to form urinary sinus.	" " "
Ceratodus.	—	Functions. Is in posterior part of cœlome. Nephrostomes present.	—	" "			Duct on ventral surface of gland ; ureters with common opening ; sperm passes out through abdominal pores ; bladder = urinogenital sinus.
Protopterus.	—	" "	—	" "			Duct on ventral surface ; ureters open separately behind genital opening and in front of rectum to cloaca ; a small urinary bladder.
Polypterus.	—	" "	—	Remains undivided.	—	—	Duct at outer margin of gland. No cloaca. Genito-urinary ducts unite.

Acipenser.	Early atrophies.	Functions; its large halves fuse anteriorly and posteriorly; anterior part degenerates. Nephrostomes present.	—	Splits only in anterior part.	Partly separated anteriorly, opens to segmental duct.	Continues into segmental duct.	Urinogenital sinus (bladder) is the dilation of hinder part of segmental duct.
Lepidosteus.	" "	" "	—	" "	" "	" "	" " "
Amia.	" "	" "	—	" "	" "	" "	" " " Urinary bladder (= a bilobed outgrowth of segmental duct) present; ureters open separately into it.
Teleosts.	" " (Functions in Fierasfer.)	Functions; halves may be very large, and partly or wholly confluent.	—	" "	" "	" "	Ureters (= segmental ducts) may be separate or united; may be separate from genital ducts; may pass to bladder, thence by short "urethra" opening behind vent. Urinary is usually behind genital opening. In some urinogenital opening passes to rectum. Ureter is sometimes in the substance of the gland.

* Probably not the metanephros in the sense of the term in Amniotes.

XIV. ABDOMINAL PORES

Cyclostomes.	Paired, on either side of vent, just within margin of cloaca (or of urinogenital sinus).
Sharks.	Paired, on either side of (somewhat behind) anus, just within margin of cloaca; absent in 4 genera of Scylliids, Cestracionts, Notidanids, Rhinids; may be absent in male and occur in female; may be absent, and reappear at spawning time (Turner).
Chimæra.	Paired, open at surface.
Ceratodus.	Paired, within cloaca, close behind anal and urinogenital openings.
Protopterus.	Sometimes paired, within cloaca, behind anus, sometimes single, when it becomes external in front of anus.
Ganoids.	Paired, on either side of (slightly behind) anus, always in front of urinogenital opening. Wanting in Amia (?).
Teleosts.	Usually wanting. Paired in Salmo. Sometimes a single pore (porus genitalis) present, opening behind vent. **In Muræna** a single pore opens into the ureter near its external opening.

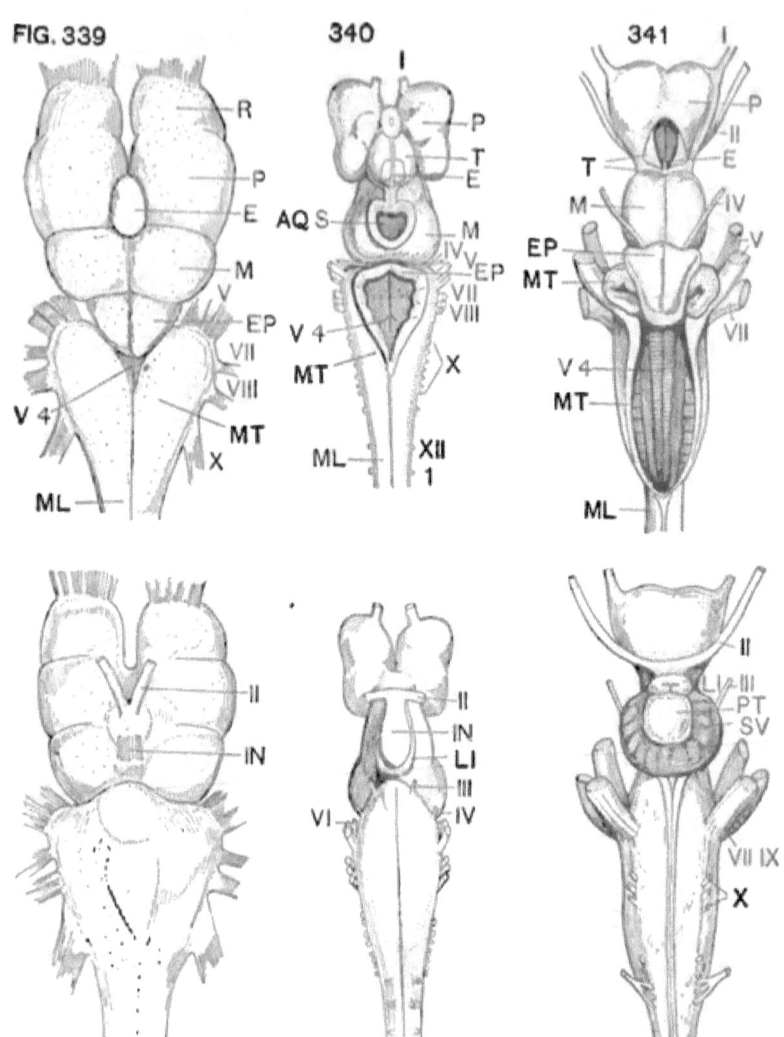

Figs. 339-344.—The brain of fishes. The dorsal view of each brain is shown in the upper figure, the ventral view immediately below. 339. *Bdellostoma.* (After JOH. MÜLLER.) 340. *Petromyzon* (*Ammocœtes* stage). (After ZIEGLER'S model.) 341. Shark (angel-fish, *Squatina*). (After DUMÉRIL.) 342. *Chimæra.* (After WILDER.) 343. Lung-fish, *Protopterus.* (After BURCKHARDT.) 344. Perch, *Perca.* (After T. J. PARKER.)

AQS. Aqueduct of Sylvius. *DSE.* Diverticula of saccus endolymphaticus. *E.*

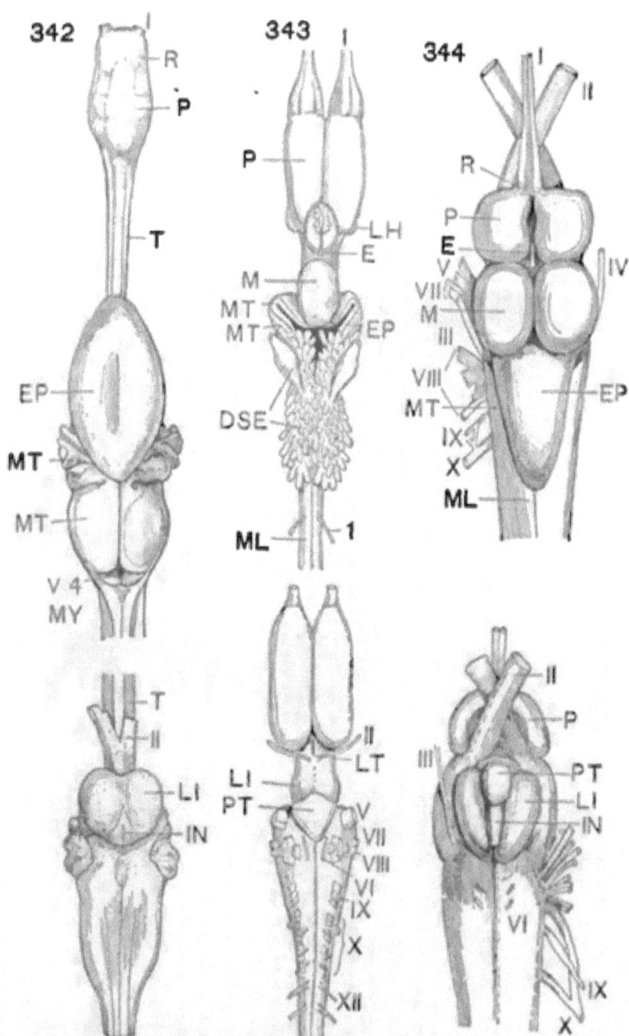

Epiphysis. *EP.* Epencephalon. *IN.* Infundibulum. *LH.* Lobus hippocampi. *LI.* Lobi inferiores. *LT.* Lamina terminalis. *M.* Mesencephalon (optic lobes). *ML.* Myelencephalon (spinal cord). *MT.* Metencephalon (medulla). *MT'.* Anterolateral lobes of metencephalon. *P.* Prosencephalon (cerebral hemispheres). *PT.* Pituitary body. *R.* Olfactory lobes. *SV.* Saccus vasculosus. *T.* Thalamencephalon. *V 4.* Fourth ventricle. Numbers *I–X.* Cranial nerves. *1.* First spinal nerve.

XV. THE CENTRAL NERVOUS SYSTEM OF FISHES

(Cf. Figs. 339–344)

A summary of the more important differences within the groups of fishes

Cyclostomes.

Bdellostoma. Brain small, compact, solid (?) ; olfactory lobes separated ; cerebral lobes (*prosencephalon*) partly separated ; the region of the *thalamencephalon* apparently greatly reduced ; epiphysis, however, prominent ; optic lobes (*mesencephalon*) and cerebellum (*epencephalon*) with median dorsal furrow ; medulla (*metencephalon*) wide and sharply tapering, separated from the brain by abrupt transverse dorsal and ventral fissures. Spinal cord flattened dorso-ventrally, its nerves with generally 2 anterior to 1 posterior root ; of these the ventral roots only unite, but dorsal branches derived from each pair of anterior roots fuse together   No sympathetic system present, unless it be regarded as represented by the long intestinal branch of the vagus.   A chiasma of the optic nerves probably occurs.

Petromyzon. Brain longer than in *B* ; its parts more readily distinguishable ; *fore-brain* with separate prominences, hemiamphicœle ; no lobi hippocampi or postolfactorius ; lobes of *thalamencephalon* raised and prominent ; a long epiphysial stalk with end organ (pineal eye ?) ; no velum (?), a simple choroid plexus, lacking inferior plexus ; a well-marked infundibulum and lobi inferiores ; *mesencephalon* with dorsal lobes and prominent central opening ; *epencephalon* small ; behind its rim a broad opening (sinus rhomboidalis) in the *metencephalon*. Spinal chord flattened as in *B* ; its nerves alternate anterior and posterior roots ; both divide into dorsal and ventral branches, but do not unite. Sympathetic system doubtful. A chiasma of the optic nerves occurs.   In the brain substance the nerve cells are scattered generally ; no cortical layers.

Fishes.

Elasmobranch. Brain elongate ; its parts well separated ; a well-marked post-olfactory lobe continues anteriorly into a tractus, hollow or solid, and then enlarges into a hollow olfactory bulb ; *prosencephalon* with ventricle more or less divided into lateral halves ; no lobus hippocampi ; *thalamencephalon* with large dorsal opening, into which a simple choroid plexus passes ; velum and a long, slender epiphysis present ; on the ventral surface an infundibulum with the paired outgrowths, lobi inferiores, — a large pituitary body, and a pad-like (paired) saccus vasculosus ; *mesencephalon* smaller than *prosen-*

cephalon; *metencephalon* with anterior lateral outgrowths immediately below a well-raised *epencephalon*; the fourth ventricle is long, shallow, and widely open. Spinal cord and spinal nerves essentially as in other fishes. A sympathetic system is present connected with the vagus. An optic chiasma occurs. The cells of the brain region are but little differentiated; ganglion cells are scattered throughout the entire mantle, with but traces of a thin cortical grouping.

Chimæroid. Brain in general shark-like, but of a remarkable form; the *thalamencephalon* greatly elongated; the *epencephalon* large, and situated high above the remaining brain parts, concealing the *mesencephalon*; the anterior lateral lobes of the *metencephalon* are greatly convoluted, its marginal rims have greatly widened, roofing over in great part the fourth ventricle. Large lobi inferiores. A chiasma of optic nerves. Spinal cord, nerves, and sympathetic system probably as in sharks.

Lung-fish. (Protopterus.) Brain elongate; fore-brain separated into halves, which terminate anteriorly in dilated olfactory lobes; lobi hippocampi present; *thalamencephalon* narrow and long, with prominent epiphysis and folded choroid plexus, below with inconspicuous lobi inferiores, and prominent pituitary body; *mesencephalon* far smaller than the fore-brain, not outwardly divided into halves; *epencephalon* (in Ceratodus?) not differentiated, suggesting the conditions of lamprey and Urodela; *metencephalon's* antro-lateral lobes similar to those of shark; posteriorly the fourth ventricle is roofed over with diverticula of the saccus endolymphaticus. The chiasma of the optic nerves faintly indicated. Spinal cord in general shark-like. A systematic system wanting (? W. N. Parker). The nerve cells of the brain are but faintly differentiated; the corpus striatum is primitively striate; and ganglion cells have been differentiated into a central grey substance and into the two cortical layers.

Teleost. The nervous system of Teleosts may be regarded as an extreme modification of that of the Ganoids, through which its structures are referable to elasmobranchian conditions. The brain greatly reduced in length, and its histological characters widely modified; the *prosencephalon* small, its olfactory lobes greatly reduced, its dorsal region distinctly lobed (lobus hippocami?); *thalamencephalon* wide, but hardly to be seen in dorsal view; its epiphysis is rudimentary, and the choroid plexus simple, slightly folded; on the ventral side a long infundibulum occurs, with greatly enlarged lobi inferiores and a prominent pituitary body; the *mesencephalon's* dorsal (optic) lobes of great size, distinctly paired; *epencephalon* prominent, overlapping and concealing posteriorly the region of the fourth ventricle. An optic chiasma does not occur. The sympathetic system connects anteriorly with the third cranial nerve (as in higher vertebrates), not with the vagus, as in sharks and some Ganoids. The substance of the brain is highly differentiated; the roof of the fore-brain is non-nervous, or with traces of differentiation of the cortical cells.

XVI. THE SENSE ORGANS

	Nasal Organ.	Eye.	Auditory Organ.
Cyclostomes.	Apparently unpaired, the capsule deeply implanted in the head immediately in front of the brain; an unpaired opening in the snout region or in the head roof, and (in Myxinoids) a median ventral opening through the mouth roof. The nerve supply of the olfactory capsule is paired.	Contains the usual elements in the adult Petromyzon; in Myxinoids lacks eye muscles, sclerotic, iris, lens; these are also lacking in Ammocœte, the overlying skin functioning as cornea. No tapetum is present.	Small and undeveloped; consists of vestibule lined with cilia, but no longer connected with the surface; to this is added a single, stout, semicircular canal in Myxinoids, and 2 (vertical) in Petromyzon; in the latter ampullæ and simple saccular and utricular outgrowths occur. The ductus endolymphaticus (Myxinoid) is reduced in length, with a bulbous tip.
Sharks.	Each olfactory capsule relatively large, cup-shaped, with stout radiating septa bearing the sensory epithelium; its nerve supply (non-medullated, as in all forms) is derived from an enormous olfactory bulb, closely adjacent. The nasal organ may or may not be connected with the upper rim of the mouth; its outer opening is subdivided into anterior and posterior portions by an overlapping (in cases fusing) marginal flap.	Capsule large, simple; sclerotic strengthened with cartilage; nictitating membrane and eyelid (dermal folds) usually present. No glands.	Vestibule simple, embedded in the head cartilage, but retaining (usually) open connection with the surface; sacculus and recessus are suggested, but undifferentiated; the 3 semicircular canals are large and well marked; ampullæ are present, but are not notably specialized.

Chimæroids.	Nasal capsule as in sharks; its nerve supply is derived from an **olfactory** bulb of greatly reduced **size**; **the nasal openings become extremely complicated by the specialized labio-nasal cartilages. The anterior naris** is somewhat tubular, **opening anteriorly close to its** opposite fellow under the snout tip; the posterior naris is beneath **the** lip margin at the juncture of palatine and "vomerine" dental plates.	Capsule **of great size, its characters** similar to those of **sharks.**	**Vestibule and its duct as in shark;** the **sacculus, recessus, and ampullæ** appear more **perfectly evolved.** Organ separated from brain by membranous wall only.
Dipnoans.	**Nasal** capsule, elongate, somewhat tubular, flattened **at** the sides, with transverse (slightly radiating) septa; its nerve is greatly reduced in size; the anterior and posterior nares **are the** openings of the tubular **capsule; the former** appears within the rim **of** the snout, the **latter** within the mouth at the inner margin **of the** "vomerine" plate; there are no dermal flaps.	**Eye small, with** feeble musculature; lens large, spherical, almost touching **the** cornea, lacking, however, in ciliary **muscles. No argentea.**	Vestibule and its duct as in sharks; recessus little developed, sacculus with bulb-like lagena. Organ separated from brain **by** membranous wall only.
Teleostomes.	**Nasal capsule shallow, flattened,** small, with **or without sensory septa;** its nerve supply is from the adjacent greatly reduced nasal bulbs; its position **is** usually on the **latero-dorsal** side of the head; the nasal openings **are separate,** small, their ectal rims often produced **tube-like.**	Eye highly specialized, capsule often bony; argentea present; a **falciform** process (pigmented, **vascular, nervous, muscular) arises at the back of the bulb, and is connected distally (by the campanula Halleri) with the** equator **of the lens; it is concerned** probably in accommodation. **A cho**roid gand usually **occurs near optic** nerve between choroid and **argentea. No tapetum is present.**	With many specialized structures; duct **no** longer open at the surface; recessus prominent; **sacculus well** constricted **off, with** elongate **often** prominent lagena. Otoliths large, widely modified. **Organ** separated from brain by membranous or **carti**laginous wall.

XVII. CHARACTERS OF INTEGUMENT AND INTEGUMENTARY SENSE ORGANS

	Skin.	*Dermal Sensory Organs.*
	In general relatively thick; clearly two-layered. The *epidermis* is composed of large and definite cellular elements, often differentiated into monocellular glands (goblet cells, retort cells); the *derma* consists of many layers of loosely arranged connective tissue cells, but is lacking in muscular elements.	
Cyclostomes.	Skin layers clearly marked, the outer highly glandular; the finely striate cuticular border is probably (Wiedersheim) referable to a ciliated condition. In Myxinoids the three outer cell layers are composed notably of goblet cells; together with these are other large granular cells, which come to the surface and discharge their contents, as thread tangles, forming the network to retain the mucous secretion. In Petromyzon the epidermis contains superficial goblet cells, elongated club-shaped cells (which rise to the surface and burst), and large granular cells, whose processes are connected with the corium; its cuticular border is pierced with fine pores.	Detached rod cells, with ciliate processes, generally distributed. In Petromyzon nerve end buds sunken in pits, are arranged in an upper and lower lateral line, and in three typical lines on the head. Myxinoids are lacking in these structures, but are remarkable in a sub-connective tissue, (fatty) nervous layer.

Sharks.	Epidermis does not present a sharply marked boundary with the derma; its elements are relatively **of** small size, generally vascular, without flask-shaped mucous cells; it lacks a stratum corneum, and is partly shed during the larval period. Shagreen.	**Lateral line and head canals usually sunken and tubular** but groove-like **in generalized** forms; simple tubular ducts to the surface. Specialized ampullary canals (Lorenzini) occur in the head region. Innervation by facial and vagus. No scattered sense buds.
Dipnoans.	Epidermis with striate cuticular border; sharply **marked off** below from derma; its elements are relatively of **great** size, loosely compacted and highly glandular; goblet **glands** are present, and, as in Amphibia, muticellular glands, dipping down into the derma. Ganoidean and cycloidal scales.	**Lateral** line and head canals similar somewhat to sharks. A groove-like condition in the head; specialized to scales in the trunk. Innervation of head region by special pre-auditory roots of facial, of trunk by special root of vago-glosso-**pharyngeus.** In addition sense buds and hillocks are very numerous **both** in head **and trunk regions.**
Chimæroids.	Skin **shark-like; without special integumentary glands.** Shagreen **and dermal plates.**	Lateral line shark-line in arrangement **and innervation (?)**; **no ampullary** organs (?); no scattered sense **buds.**
Teleostomes.	Epidermis of well-marked polygonal cells, with striated **cuticular** border; sharply marked off below from derma; contains plentiful goblet mucous **cells,** which discharge their secretion at the surface; **branched cells** plentiful, pigmented **and non-**pigmented. Derma **of widely differentiated** fibrous layers. Ganoidean or horn-like scales.	Sensory canals shark-like in arrangement, **especially in Crossopterygian;** external openings minutely **subdivided (cluster pores)** in Ganoids, and adapted **to horn-like scales in Teleosts; organs** innervated as in **Dipnoan (with the addition, perhaps, of trigeminus,** Collinge). **Scattered sensory organs (pit organs, Merkel's buds, smell buds) especially numerous in Teleosts.**

XVIII. THE EARLY DEVELOPMENT OF FISHES

	Marsipobranchs.		Shark.	Chimæroid.	Dipnoan.	Ganoid.	Teleost.
	Myxinoid.	*Lamprey.*					
Size of eggs (millimetres).	18.	1.	50–150.	140.	3–5.	2–3.	1–10.
Membranes.	Long, horny capsule, with terminal processes.	Soft membranes, the outer distending but little.	Horny capsules with stout processes. Membranes soft in Læmargus.	Horny capsule with fringing thread-like processes.	Soft membranes, the outer greatly distensible in water.	In Acipenser apparently as in lung-fish. In Lepidosteus somewhat as in lamprey.	As in Ganoids, but with many variations, sometimes greatly distensible, sometimes with horny capsule and processes.
Fertilization.	In abdominal cavity (?).	After extrusion.	In oviduct; in Læmargus after extrusion.	In oviduct; in Harriotta after extrusion.	After extrusion.	After extrusion.	After extrusion usually.
Number of eggs.	30 (?).	30,000. Ripen simultaneously.	4–12 (?), others ripening afterward.	2 (?).	Several thousands, ripening at various times during spawning season.	7,000,000 (sturgeon); ripen simultaneously; many spoil before being deposited.	50–10,000,000; usually ripen simultaneously.
Length of time in hatching.	Months (?).	1–2 weeks	8–15 months.	Months (?).	1–2 weeks.	1–2 weeks.	5 days–3 months.
Segmentation.	Meroblastic (?).	Holoblastic.	Extremely meroblastic.	Meroblastic.	Holoblastic with early variations.	Usually holoblastic; early stages of Lepidosteus and of Amia somewhat meroblastic.	Extremely meroblastic.
Blastula.	——	Large segmentation cavity, with two-celled roof.	Flattened segmentation cavity, whose roof is many cells thick, whose solid floor is filled with merocytes.	Shark-like probably.	Somewhat like Ganoid (Acipenser).	Segmentation cavity slightly flattened; its roof of several cell tiers; its floor either cellular or filled with merocytes.	Greatly flattened segmentation cavity.

Gastrula.	—	Gastrula readily comparable with that of Amphioxus; coelenteron is wide, but pouch-shaped; ventral lip of blastopore swollen with yolk.	Coelenteron greatly flattened; the blastoderm a flattened cell mass, with slightly thickened rim, lying on the yolk (= ventral wall of coelenteron).	—	Similar probably to that of Ganoid or Amphibian.	Invagination begins at dorsal lip of blastopore, extends thence around the lower hemisphere; coelenteron is thus a fissure between the round mass of yolk cells and the blastoderm; is deepest in dorsal region.	Gastrula obscure; similar to Ganoid, admitting that more perfect relations of yolk to blastoderm (by means of periblast), permit the fissure-like coelenteron to merge with the segmentation cavity.
Blastopore.	—	The typical pore-like opening of the coelenteron.	The blastopore is represented by a nick in the dorsal margin of the blastoderm; the blastoderm's entire rim is the circumcrescence margin. A neurenteric canal occurs.	—	Similar to Ganoid. Neurenteric canal occurs.	Blastopore probably the indentation of the dorsal lip; the circular opening with yolk plug corresponds to the circumcrescence margin; neurenteric canal in Acipenser.	The indentation of the dorsal lip does not occur; the thickening of the tail region is here the closed blastopore; the rim of the blastoderm (germ ring) is the circumcrescence margin. No neurentric canal.
Mode of form-growth of embryo.	—	Elongation of entire gastrula, especially of anterior end.	Embryo appears pressed upward from the rim of the blastoderm; as the embryo rises above the egg, the tail grows backward, leaving a stalk connecting yolk and embryo.	—	Similar somewhat to Ganoid, lamprey, and Amphibian; the fore part of the body elongates, leaving the yolk in the hinder ventral region.	Early rises above the yolk, shark-like, outlining sharply head, trunk, and tail (Lepidosteus), but has not the yolk stalk. Yolk sac in pectoral region.	Similar somewhat to Ganoid; large yolk sac in pectoral region.
Mesoblast.	—	Arises somewhat as in Amphioxus, suggests archenteric pouches	Solid outgrowth of entoderm on either side of hinder median line.	—	—	As in sharks; also largely peristomal.	Of obscure origin; probably Ganoidean.
Central nervous system.	—	As a solid ectodermic cord dips down into embryo's axis; later acquires lumen.	A thickened axial plate, whose margins arch up and fuse in the median line, closing from the mid-region forward and backward.	—	Neural folds rise above the surface, arch over, and fuse in median line; closure as in sharks.	Similar to the conditions of sharks and Dipnoi.	Similar somewhat to lamprey.
Paired fins	—	—	As lateral dermal folds, or lappet-like outgrowths, with metameral supports.	—	Appear very late (after several months), as archipterygia.	Occur somewhat as in sharks.	Occur as in Ganoids, but show an abbreviated development.

XIX. THE SUPPOSED DESCENT

Interrelationships and lines of descent as suggested by a number of
noted on each scheme. * Denotes that the diagram is the present writer's

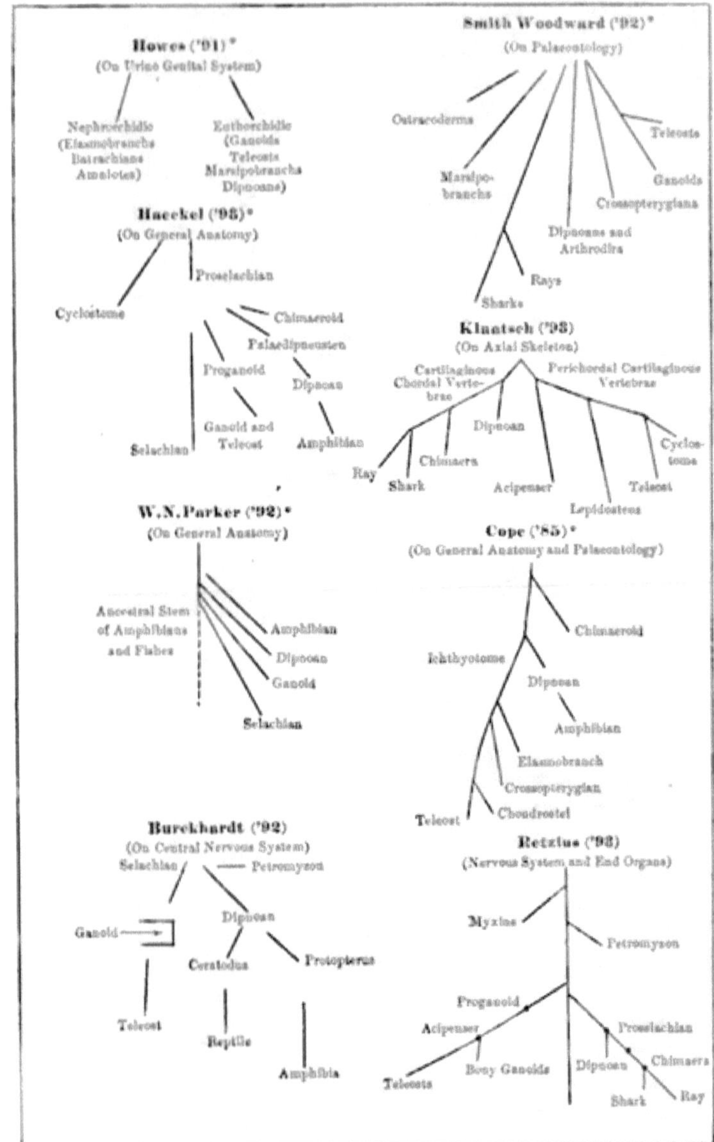

OF THE GROUPS **OF FISHES**

observers ; their views have been based on the different lines of investigation interpretation of the text of the author **cited.**

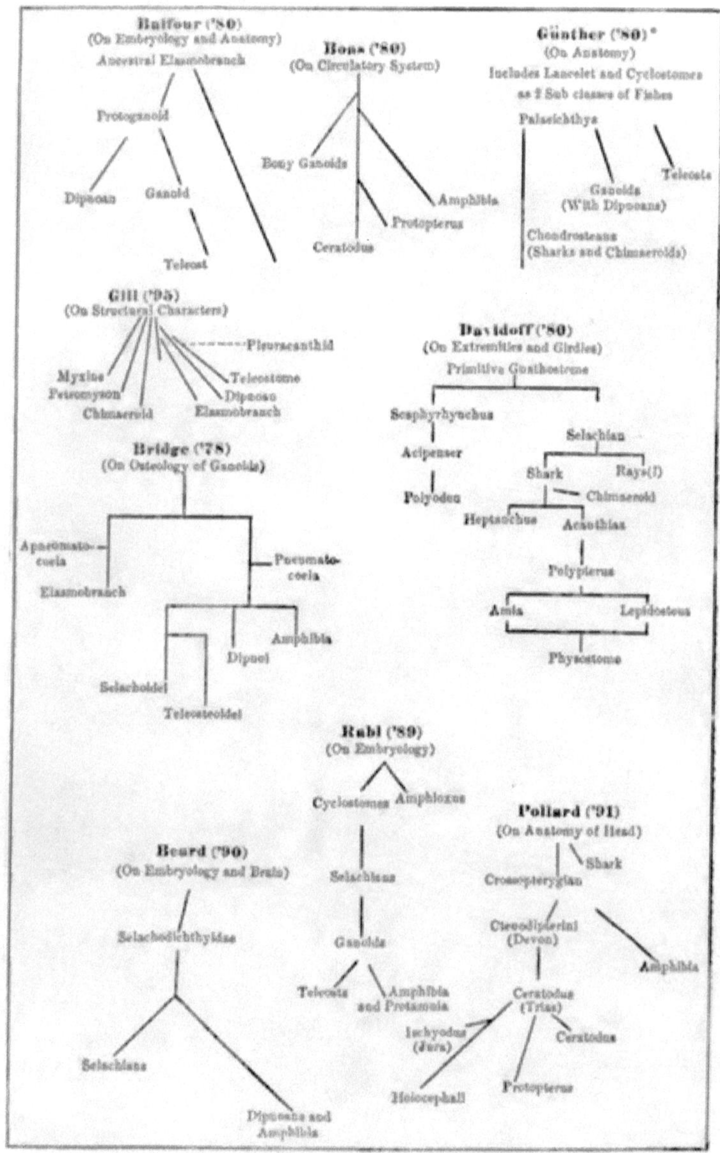

MARSIPOBRANCHS, v. Cyclostomes; tables of the early development of, 280, 281.

McClure, **182.**

Mechanical **adaptation of the** fish's **form, 5,** 6.

Median fins, v. Fins.

Megalurus, 165; *M. elegantissimus,* 165 (Fig. **171).**

Megaptera, v. Whale; *M. longimana,* numerical lines of, 5 (Fig. **7),** 61.

Menaspis, skin defences of, 113.

Menhaden, v. *Brevootia.*

Metamerism, vertebrate, of fishes, 14–16; of lampreys, 15; of sharks, 16.

Miall, L., **126.**

Microdon, **157;** *M.* **wagneri,** 158 (Fig. 163).

Mivart, St. G., **40**

Modern fishes, v. Teleostomes.

Mollier, S., **39.**

Monk-fish, v. *Rhina.*

Mormyrus, 171, 172; *M. oxyrhynchus,* 172 (Fig. 179).

Morphology, Journal of, quoted, 51 note.

Mouth of fishes, v. Jaws, Teeth, etc.; of catfish (a Teleostome), 64 note.

Movement in water, 1, 2 (Figs. 1 and **2).**

Mucous canal system, v. **Lateral line.**

Müller, Johannes, **145.**

Mullet, gills of, **20.**

Muræna, 173.

Myliobatis, v. Eagle-ray.

Mylostomids, in classification, 8; trunk of, 136; jaws of *Mylostoma variabilis,* 136, 137 (Fig. 138).

Myriacanthus, in classification, 8; restoration of, 104; head region of, 105 (Fig. 106); dermal plates of head and snout, 105 (Figs. 106, A and B), 113; mandibular, 106 (Fig. 107); dorsal spine, 107 (Fig. 114); dental evolution, 112; shagreen tubercles and dermal bones and plates, 105 (Fig. 106), 107 (Fig. 114), 113.

Myxine, classification, 8; gills of, **17** (Fig. 10), 18; general description, of *M. glutinosa,* 59, 60 (Fig. 71), **61** (Fig. 72 *B*); eggs of, 180–182 **(Fig.** 187); genital system, tables of, **266;** excretory system and urinogenital ducts, 270.

Myxinoid, Californian, gills of, 18; teeth of, 57; eggs of, 182 (Figs. 186 *A* and 187 *A*); comparison tables of the early development, 280, 281.

Names, list **of authors** and **their** works, 231–251.

Names, list of derivations of, 227–230.

Nares, in Dipnoans, **129.**

Natterer, J., 125.

Necturus, swim-bladder of, 21.

Nervous system, central, 272 (Figs. 339–341), **273** (Figs. 342–344), **274, 275.**

Newberry, J. W., **78,** 106, 120, 130, **131, 132, 136.**

Newton, **106.**

Nicholson, H. A., 125.

Notacanthus sexspinis, 168 (Fig. 174).

Notidanus, antiquity of, 9; gill slits, 16, 19; pectoral fin, 40–42 (Fig. 52), 44, 45; described, 87–89 (Fig. 93); affinities, 96; skull, jaws, and branchial arches of, 254 (Fig. **311).**

Numerical lines of fishes, 5, 6 (Figs. 5–8).

Onychodus, in classification, 8.

Operculum of Teleosts, 19; comparison tables of, **260.**

Ophidium, barbels of, **46,** 47 (Fig. 55).

Opisthure, 111.

Osteolepis, in classification, 8; description of, 150, 151 (Fig. 152).

OSTRACODERMS, classified, 8; antiquity of, 9; description of, 65–71; types of, 67; affinities of, 66 (Fig. 77), 70; list of authors and works on Ostracoderms, 238.